Hellenistic and Roman Naval Wars
336–31 BC

Hellenistic and Roman Naval Wars

336–31 BC

John D. Grainger

Pen & Sword
MARITIME

First published in Great Britain in 2011
and republished in this format in 2020 by
Pen & Sword Maritime
an imprint of
Pen & Sword Books Ltd
Yorkshire – Philadelphia

ISBN 978-1-52678-232-8

Typeset in 11pt Ehrhardt by Mac Style

Printed and bound in the UK by CPI Group (UK) Ltd, Croydon, CR0 4YY

Pen & Sword Books Limited incorporates the imprints of Atlas, Archaeology,
Aviation, Discovery, Family History, Fiction, History, Maritime, Military,
Military Classics, Politics, Select, Transport, True Crime, Air World, Frontline
Publishing, Leo Cooper, Remember When, Seaforth Publishing, The Praetorian
Press, Wharncliffe Local History, Wharncliffe Transport, Wharncliffe True
Crime and White Owl.

For a complete list of Pen & Sword titles please contact
PEN & SWORD BOOKS LIMITED
47 Church Street, Barnsley, South Yorkshire, S70 2AS, England
E-mail: enquiries@pen-and-sword.co.uk
Website: www.pen-and-sword.co.uk

Or
PEN AND SWORD BOOKS
1950 Lawrence Rd, Havertown, PA 19083, USA
E-mail: Uspen-and-sword@casematepublishers.com
Website: www.penandswordbooks.com

Contents

List of Maps and Illustrations

Maps

Illustrations

All images are from the author's collection.

1. Warship in mosaic showing Odysseus and the Sirens, Tunisia
2. Warship in mosaic showing fight with sea monsters, Tunisia
3. (a) The *Isis*, fresco from Nymphaion in the Crimea
   (b) Drawing of the *Isis* fresco
4. A Roman trireme, carved panel from Puteoli
5. Ship construction, remains of a Roman vessel from Mainz
6. 'Navis tetreris longa', drawing of a quadrireme from Alba Fucens
7. Stern of Rhodian galley, drawing of relief from Lindos
8. Quinquereme in action, mosaic from Piazza Barberini, Rome
9. Quinquereme, artist's reconstruction from sculpture at Isola Tiberiana, Rome
10. Roman warship and marines, relief from Naples
11. Roman warship and marines, relief from Rome
12. Antiochos III
13. Coin of Demetrios Poliorketes showing prow of warship
14. Coin of Antigonos Doson showing prow of warship
15. Olympos, Lykia
16. Gabala (Jeble), Syria
17. Bay of Haifa
18. Seleukeia-in-Pieria, Syria
19. Korakesion (Alanya), Pamphylia, view over the harbour
20. Korakesion, view of shipsheds and fortified hill behind
21. Phaselis, Lykia
22. Eryx, Sicily
23. Carthage, view from 'Admiralty Island'
24. Carthage, aerial view of the ancient harbours

Maps

Map 1: The Ptolemaic Empire and the Eastern Mediterranean.

Map 2: The Aegean Sea.

Map 3: Ionia-Aeolis – The Battle Zone.

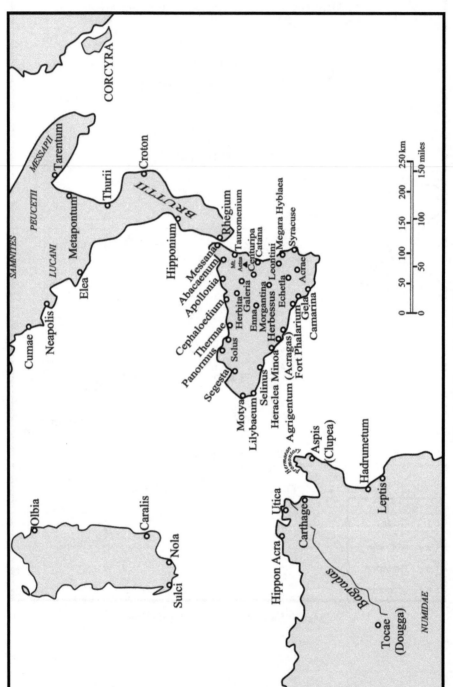

Map 4: Sicily and its Neighbours.

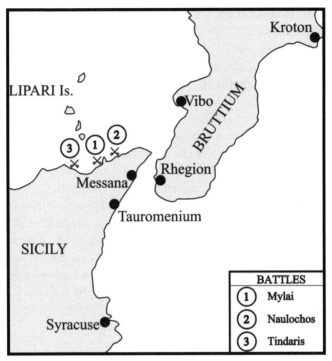

Map 5: The Strait of Messina.

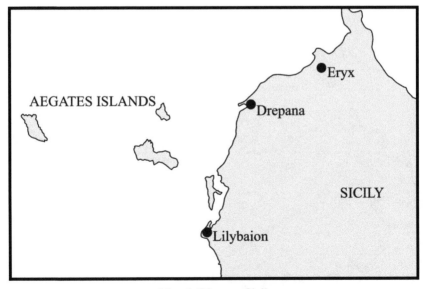

Map 6: Western Sicily.

The Great Harbours (Maps 7A, 7B and 7C)

In the Hellenistic world there were four particularly notable harbours. The original in many ways was Peiraios, the port of Athens (Map 7A), where the natural conformation of the shore provided a small harbour, Zea, which was used as a naval port, while the larger bay, Kantharos, was the commercial base. Syracuse (Map 7B) was similarly natural, where the island Ortygia separated the main harbour from the smaller naval harbour ('Laccius'). Ortygia had been an island, so originally there was a clear communication between the two.

The other two great ports were artificial. Carthage (see plate xx) was excavated from the mud of the shore. There was a rectangular commercial harbour, with long wharves for unloading ships lying lengthwise to the quay, and a circular naval harbour which had an administrative island in the centre. Each harbour had a separate entrance from the sea. Alexandria (Map 7C), laid out in the last years of Alexander and the beginning of the rule of Ptolemy I, was built along with the city, and made use of the island of Pharos. This was connected to the mainland by a mole, so forming two harbours. Each of these had a section which was official – for the Palace and the Kibotos harbour – and there was a third harbour on the shore of Lake Mareotis, connecting with the River Nile, along which came the supplies for the city.

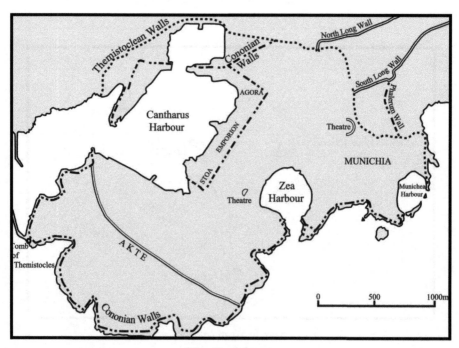

Map 7a: Athens.

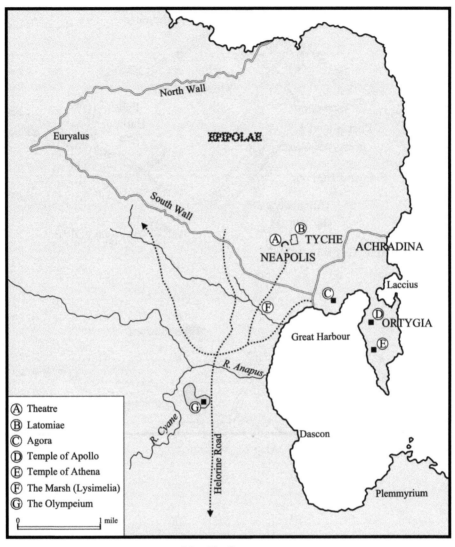

Map 7b: Syracuse.

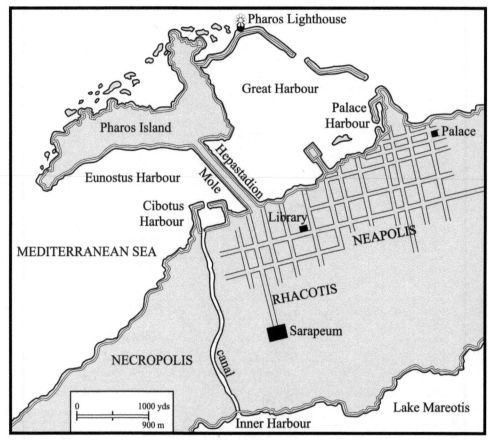

Map 7c: Alexandria.

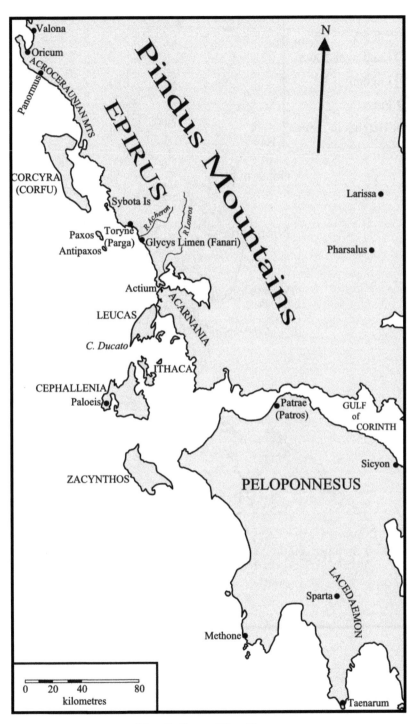

Map 8: Greece West Coast.

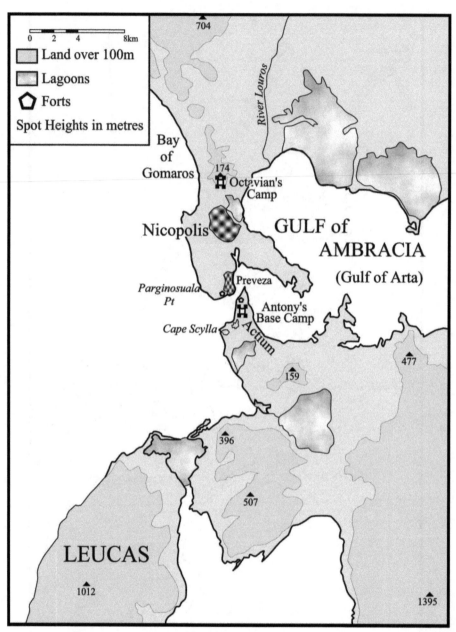

Map 9: Actium.

Introduction

Between the expeditions of Alexander into the Persian Empire starting in 334 BC and that of Octavian (later to be Augustus) to Egypt in 30 BC, the Mediterranean was repeatedly the scene of major warfare at sea. This was not a new thing, of course, since several cities and states around the Middle Sea had indulged in war at sea in the past, but in the Hellenistic period it occurred more frequently, over a wider range, and with greater intensity than before – or than at any time afterwards until the great wars of the sixteenth century AD.

This was a period which saw several states deliberately building up navies to serve as major instruments of power. There was competitive pressure on all sea-states to build more, bigger, and better ships – a naval arms race the like of which was perhaps not seen until the early years of the twentieth century AD. The competition involved not just building and crewing more ships than any rival, but also innovations in size of ships, and in the technique, tactics, and strategy of war. Different methods of naval warfare developed and were tested in battle, and navies were deliberately used to impress and intimidate potential allies and enemies.

Such competition was expensive, not only in the building of ships and in paying and feeding their crews, but also in constructing shore installations. Harbours large and safe enough to hold great fleets had to be built and facilities to hold supplies of naval tackle ready to build new ships and repair those damaged had to be gathered and stored. Ships had to be maintained and repaired and replaced once they wore out or became rotten. And in battle a sunken ship took with it not only the investment made in its construction and use, but usually most of its crew as well.

Lost battles at sea were capable of stopping an advancing empire in its tracks. Victories at sea persuaded men to declare themselves kings. The acquisition of power at sea was one of the major elements in the overall political contest in the Mediterranean world, which involved half a dozen great powers and dozens of small ones. This contest was won in the end by Rome, of course, by the early years of the second century BC, and it was by the use of sea power that Rome prevailed. This is in a sense paradoxical, since Rome is generally regarded above all as a successful land power – the legions, and so on. Nevertheless it was only by the use of its naval power that Rome won its wars. Think of Italy's geographical situation: only by sea could a Roman force reach any of its enemies.

Sea power in the Hellenistic period, therefore, is at the heart of the political conflicts of the time. It is perhaps an even greater element in the success or otherwise of the various states in that context than any other aspect of power. Its creation, with its ships, harbours, and crews, was a central part of the development of the power of the state in that time as well, and the expenditure involved was a major element in taxation and finance, in the employment of artisans, and so in economic development generally. The end result was the overwhelming state power of the Roman Empire, control of which was won by Octavian/Augustus in a sea battle. After his victory at Actium most of the navy was dismantled. With no enemy at sea, a navy became an expensive and unnecessary luxury.

Chapter 1

Alexander's Naval War

When Alexander of Macedon set off to campaign against the Persian Great King in 334 BC, he had no experience of war at sea.* His father, King Philip II, whose career of intrigue and victory was his inspiration, had built up a small navy, and had used it in a characteristically cunning way. Both of them faced Athens and its large fleet of warships, and both, in attacking Persia, were tackling the other great naval power of the Mediterranean. One of Alexander's priorities was to keep these two apart, since if they combined against him their fleets would make every part of his home kingdom vulnerable. For several years he was careful to hold Athenian hostages and at the same time to pay court to Athenian sensibilities.

Alexander's invasion force consisted of a formidable army and a weak fleet. Also he had very little cash at first and, like most conquerors from Sargon of Akkad to Adolf Hitler, he aimed to finance his expedition out of the booty of his conquests. Alexander's fleet consisted of 160 triremes, of which an unknown number were Macedonian; probably most were contributed by the Greek allies of the League of Corinth, though the only specific contribution known is the twenty ships of Athens.[1] It seems probable that the burden was spread widely amongst the allies, partly for financial reasons – the allies were paying their own men – and partly to forge a wider sense of participation in the campaign.

In getting his army across the Hellespont and into Asia these warships were assisted by an unspecified number of cargo ships, presumably hired or impressed for the occasion. No Persian ships were able to interfere with this crucial initial movement. Indeed it is not clear that anyone on the Persian side realized what was happening, for no Persian mobilization had yet begun. Once Alexander was victorious over the local Persian forces in western Asia Minor in the battle at the Granikos River in 334, he moved south to liberate the Greek cities of the Aegean coast and the fleet moved south in company.

The Persian reaction to this early setback was to organize the assembly of a great army, and to gather a large fleet from the Phoenician, Cypriot, and

*Hereafter, all dates are BC, unless specified otherwise.

Kilikian cities. Some of these vessels had been used in 335 to suppress an Egyptian rebellion, and the construction of more vessels was commissioned. The figure of 400 ships is mentioned, but it is unlikely that this was ever achieved, and another figure of 300 is reported later.[2] There were fleets at the cities ready to be mobilized and used, but it would clearly take some time to build a new set of ships. So the Persian ships which confronted Alexander's fleet at Miletos in the summer of 334 were only the first contingent. There were certainly not 400 ships, for Parmenion, Alexander's cautious senior general, is said to have recommended that the Persian fleet be attacked. He would not have expected 160 Greek and Macedonian ships – and some of the Greeks of suspect loyalty – to beat a Persian fleet more than twice that size. So we must assume the Persian fleet was roughly the same size as Alexander's at the time, perhaps 200 or less ships, but with more ships likely to turn up at any time. The implication of an approximately comparable strength is encouraged by the fact that neither side was willing to engage the other in battle. The Greek-Macedonian fleet merely ensured that local water points on the coast were occupied to deny them to the enemy, then blocked access to Miletos harbour during the army's siege of the city. The Persian fleet retired to Mykale several miles away, and then had to get supplies from Samos, which was an Athenian island – a good indication of the anti-Macedonian attitude of at least some of the Greeks.[3]

The lesson Alexander learned was that a fleet could be kept at bay by controlling its land supplies. His ships were costing him a good deal of money. Both 100 and 250 talents a month have been suggested, and the Greek allies seem to have failed to contribute their 'share' of the expenses. Except for the Hellespont crossing, the fleet had been of little use. Further, if the Persian project of a fleet of 400 ships was reached, his own fleet would be overwhelmed, and he could not afford a serious defeat so early in the campaign. So Alexander disbanded his fleet, except for the twenty ships of Athens and a few others, and some of his own Macedonian ships. The Athenian ships and crews could be seen as hostages for the city's good behaviour.[4]

The decision to disband the fleet was soon shown to be premature, since a substantial fleet would have been very useful at the siege of Halikarnassos which followed in the autumn. Skilfully used, this could have deterred the Persian fleet. Alexander could have appealed for more ships from the League members, principally Athens, which had a fleet of over 400 ships available. That would, however, have tied him to the Aegean area until the Persian fleet was driven off, and would also have made him hostage to Athens, the reverse of the position he wished to be in. So, if the fleet was costing him a lot, and was liable soon to be beaten anyway, he might as well collect the soldiers from it and send the ships home. The only other use for the ships was to transport supplies, and by then he controlled enough resources in Asia to live off the land. So he marched off to the east to attack the interior of Asia Minor. He will have understood, even if he did not have accurate information about it, that a new Persian Army was being collected in the east. It was clearly best to meet it away from the vulnerable Aegean coast.

The Persian strategy in Asia Minor, implemented by the Greek mercenary general Memnon of Rhodes, was to carry the war, by means of the fleet, into Alexander's home territory, and to encourage disaffection and perhaps revolt among his allies. The fleet had a free hand in the Aegean, for a time. To be effective it had to range widely and seize the initiative, but it also required a secure local base. Memnon began well, gaining control of Chios to add to Kos, which was already under the fleet's control, and these islands gave him adequate bases. He could count on Samos to be at least benevolently neutral, for its Athenian colonists had already helped his fleet when it needed supplies during the fight for Miletos. From Chios he moved north to Lesbos, won over several of the cities of that island, but then sat down to besiege Mitylene. Most of the Kyklades Islands sent envoys to him. This was not quite submission to Persia, but it was certainly the first step on the way.[5]

Memnon was thus slowed in his progress by the need to take Mytilene, but it was sufficiently disturbing for Alexander to reverse his non-naval policy. From his winter quarters at Gordion, in central Asia Minor, he sent two of his officers, Hegelochos and Amphoteros, with 500 talents to revive the fleet. There had been time enough, evidently, to return the ships to their home ports when the fleet was disbanded and to lay them up for the winter of 334/333; reassembling and crewing them again would take some time, but the Macedonian-Greek fleet took the sea again in the spring of 333. Hegelochos was to command an army to guard the Hellespont, now seen to be a vital element in Alexander's communications; and Amphoteros was to defend the islands, and to take command of the fleet. In addition, Antipater, Alexander's viceroy in Macedon, was sent 600 talents, presumably to assist in the re-equipping and crewing of the ships. The league members were also to provide ships, no doubt those they had received back the previous autumn.[6]

It is curious, to say the least, that within a few months of disbanding the fleet partly because it was too costly, Alexander was able to send over a thousand talents home. The money obviously came from the cities and lands he had already conquered and the Persian treasury at Sardis, and it seems difficult to believe that he did not have access to adequate supplies of money earlier – when the decision was made he already held Sardis and most of the cities of the Aegean coast, all of them fruitful sources of taxation. This rather suggests that the real reason for disbanding the fleet was that it was too small to face the Persian force.

Arrian, Alexander's biographer, tells a tale whereby Alexander interpreted the sighting of an eagle on shore astern of the Macedonian fleet to mean that he would be able to defeat the Persian fleet by controlling the land.[7] If this is so, he then changed his mind (and so no doubt the omen was conveniently forgotten or re-interpreted). Further, it was not a strategy he put into practice, for to do so he would have needed to garrison every port, and to patrol every river mouth. He did not do this, any more than he garrisoned every place inland. It is best to accept, as Alexander eventually did, that disbanding the fleet had been a bad mistake. It is all to his credit that he reversed himself when it was seen to be necessary.

It was going to take time to reassemble the fleet, and had Memnon been able to get on with his own campaign it might well have been too late. The obvious Persian strategy was to encourage Greek defections, especially among the mainland cities in Greece. This could threaten the Macedonian homeland, and, if Antipater, Alexander's viceroy and commander in Macedon, could be beaten, this ought to bring Alexander home hot-foot. The fleet had made some progress in this under Memnon's command, but then suffered two disabling blows. First Memnon was ordered to send all his Greek mercenaries to join the new Persian army which was being prepared to meet Alexander in battle. They went off under the command of Memnon's nephew Thymondas in a large fraction of his fleet, perhaps 200 ships.[8]

Then, in about June 333, Memnon himself died. He appointed his nephew Pharnabazos to take over the command, and Pharnabazos was competent enough, but inevitably the campaign stalled further. Mitylene was finally taken (perhaps even before Memnon died), and Pharnabazos was able to threaten the Hellespont by taking Tenedos, but it is likely that by then Hegelochos had beefed up the garrisons in the many small cities along the Hellespontine coasts, though Amphoteros had not yet been able to collect enough ships to be able to challenge Pharnabazos' fleet. A Persian squadron which penetrated the Hellespont was defeated, and either then, or on the removal of the Persian main fleet, the Tenedians revolted from Persian control.[9]

Another detachment of ten ships, commanded by Datames, had been sent south from Mitylene to Siphnos, one of the Kykladic islands which had presumably sent envoys to Memnon earlier. No precise purpose is stated for the move, but it must have been in connection with the preliminary moves by King Agis of Sparta, who was showing clear signs of responding to Persian promptings. Sparta was not a member of the League of Corinth, so technically Agis could have diplomatic relations with the Persians without falling foul of Antipater and Alexander, though even by listening to Pharnabazos or Datames he was showing hostility to Macedonian pretensions. It may also have been a response to the Kykladic contacts initiated by Memnon. Ten ships would not be enough to give serious help to Agis, but they may have encouraged Persian sympathizers in Siphnos and the Kyklades, and they were adequate as an escort for a diplomatic mission. Combining two tasks would make sense to Pharnabazos.[10]

Alexander's instructions to the league to collect a new fleet had begun to have an effect by this time. A Macedonian, Proteas son of Andromachos, had command of a squadron of fifteen ships at Chalkis in Euboia. Hearing of Datames' arrival at Siphnos he sailed to nearby Kythnos and, in a dawn attack after a careful reconnaissance, he captured eight of Datames' ships; the remaining two, with Datames himself, escaped. Proteas was clearly able to gather information easily, and no one on Kythnos warned Datames, while word had gone from Siphnos to Proteas at Chalkis very quickly. It seems clear that, even if some of the Kykladians sent envoys to Memnon, many others – Arrian says most – remained sympathetic to Alexander.

Pharnabazos had used the bulk of his fleet, a hundred ships, at Lesbos and Mitylene. At some point he was able to retake the city of Miletos, and the Persian garrisons in the forts at Halikarnassos seem to have been able to recover control of the city – the Persian base at Kos was close by and could give support.[11] Despite the minor setback to Datames at Siphnos, therefore, it could be said that Pharnabazos' fleet was still making some progress. Neither Hegelochos nor Amphoteros was apparently yet in a position to challenge the Persian fleet, even weakened as it was, and the small detachment under Proteas at Chalkis rather suggests that the Macedonian fleet was scattered in order to block raids and show the flag in Greece and the islands. All concerned, however, must have been listening hard for news from the east as Alexander marched his army ever further away, and as Dareios gathered his new army. Pharnabazos sent some ships to Kos and Halikarnassos, and then sailed with a hundred ships to Siphnos in order to resume negotiations with Agis, who in turn had already sent envoys to meet the Great King in Asia. News came at last from the east while Agis and Pharnabazos were talking. The Great King had been decisively beaten by the Macedonian army at Issos.[12]

At once the conference at Siphnos ended. Agis was given thirty talents and ten of the Persian triremes to help him prepare for war, if that was what he intended. The rest of the Persian fleet now broke up: Pharnabazos took twelve ships to Chios to hold that island; his co-commander Autophradates took the rest, now 75 ships, to join the squadron at Halikarnassos, no doubt to wait for news and instructions. The total fleet's size is not clear, though it was still over a hundred ships. But it was composed of squadrons from the several cities of Cyprus and Phoenicia, and those cities were now menaced by Alexander's victorious army. At least one of the Phoenician contingents, from Arados, was commanded by the city's king, Gerostratos. These contingents now sailed for home and they were followed by the Cypriot contingents.[13]

Pharnabazos was left with only a small force, the twelve ships he took to Chios, plus any already there and the few others which did not desert. Meanwhile Hegelochos and Amphoteros had finally collected their fleet together, which is said by one source to be of 160 ships. This is possible, though it is suspiciously the same size as Alexander's original fleet. Having the necessary numerical superiority at last, Amphoteros set about recovering Pharnabazos' conquests. Tenedos (if not earlier), Lesbos, and Chios were taken, and then Kos, where Pharnabazos was captured, though he succeeded in escaping. Miletos changed hands yet again, and the forts at Halikarnassos at last surrendered.[14]

But all was not yet finished. A large force of Greek mercenaries had escaped from the battlefield at Issos. They reached the coast at Tripolis, a relatively recently-founded Phoenician city (with a Greek name) where they found ships, presumably some of those which had carried Memnon's troops to Syria. Part of the group sailed off to Egypt, where their leader tried to make himself the imperial satrap, though he failed. Another group sailed west, heading for Greece. They stopped in Cyprus, but the Cypriot kings, whose fleets had now broken away from Pharnabazos' fleet, were in the process of switching sides. The refugee soldiers moved on, no doubt to the Cypriots' relief. Crete was their next stop, a place where their military skills were useful, and where Agis had been recruiting allies and soldiers. They were soon recruited by Agis.[15]

These soldiers in effect helped to revive the Persian Aegean campaign. They may well have included some of those who had originally campaigned with Memnon – one of their commanders was Thymondas – and they were all clearly very determined opponents of Alexander. An alliance between these men and Agis, to which would be added the remnants of Pharnabazos' fleet, which was still in the Aegean, would pose a major threat to the Macedonian position in the region, particularly while Alexander himself was stalled by the seven-month siege of Tyre. However, during that siege Alexander's maritime position changed, so he was essentially able now to make a serious effort by sea in the Aegean area to counter this new situation.

Alexander moved south from the battlefield of Issos, first to Arados, an island city with a mainland suburb, Marathos. The king's son, left in charge while his father was with the fleet in the Aegean, promised to switch sides, and communicated with his father. The kings of Arados, Byblos, and Sidon, and the Cypriot kings, then brought their persons and their ships to Phoenicia, where Alexander was balked at Tyre, the only Phoenician city to defy him. The presence of the Phoenician and Cypriot ships allowed a strict blockade to be imposed on the island city, and this hostility allowed Alexander to concentrate on the siege without having to worry about other local powers.[16] So the balance of naval power had now swung decisively in his favour in Syria, while Hegelochos and Amphoteros mopped up in the Aegean, though this process took even longer than the campaign in Syria.

At Tyre Alexander collected fleets from the three Cypriot cities which had them, Salamis, Amathos, and Kourion (this last a Phoenician city), from the three Phoenician cities which had come over to him, Sidon, Arados, and Byblos. He was joined by small flotillas from Rhodes (finally jumping off the fence), and

Lykia – these were thus no longer threatened by the Persian fleet, and since they had not joined Alexander earlier, their new presence may well have been intended to make up for this earlier lapse. Mallos in Kilikia also sent a small number of ships.[17]

Alexander's siege method at Tyre was to construct a mole by which he could reach the island the city was built on – a mode of attack which clearly demonstrates his non-maritime mindset. The city's two harbours had therefore to be blocked by his new fleet. The ships were sorted into two squadrons: the Cypriots' 120 ships on the north; the rest, probably more than 120 ships, on the south. He was also joined by Proteas, the cunning victor of Siphnos, in one of his own Macedonian ships. Proteas could bring him up to date on the situation in the Aegean.

Alexander's ships at Tyre were blockading the two harbours most of the time. They had to fight one major battle, when the Tyrians, who had eighty ships, came out to attack the fleet on the north, that of the Cypriot kings. They did so quickly enough and secretly enough to sink the royal quinqueremes of the three kings and drive some of their triremes on shore as well. Alexander led his other (Phoenician-Greek) fleet out from the southern anchorage and round the island to attack the Tyrians from that side, either while still fighting or when they turned back to return to the harbour. He was largely unsuccessful, catching only two Tyrian ships, whose crews escaped by jumping into the water and swimming away.[18] Nevertheless the Tyrians did not come out again, except in ones and twos. When the city was taken, its population was murdered or enslaved, and any surviving ships became Alexander's. Once the city had been conquered, therefore, he had no more worries about the Phoenicians' naval strength – it was now all his.

He marched to Gaza, which had to be besieged and taken, and on into Egypt, with his fleet pacing him along the coast, a familiar journey for the Phoenicians. Hegelochos reported to him in Egypt on his success in recovering the Persian conquests in the Aegean, but also the development of the Cretan-Agis problem, which was becoming increasingly menacing.[19] Alexander spent the winter in Egypt, during which time further developments in the Aegean were reported to him, though we do not know what they were, only that they centred on Agis. A squadron of thirty ships was put under the command of Polemon son of Theramenes and stationed to protect the mouths of the Nile.[20] And Alexander arranged the foundation of his great new city, Alexandria, at the mouth of the Naukratic branch of the river, a city which was to have a great harbour and a great future.

Alexander marched north again in the spring of 331, and when he stopped at Tyre he decided that the position in Greece and the Aegean was too dangerous to be left for Antipater to deal with. There was a rebellion in Thrace, the Spartan king was increasingly hostile, the refugee mercenaries and Agis had come to dominate Crete, and he himself was now faced with a revived Persian army which had been gathered in Mesopotamia by the Great King. So at Tyre Alexander commissioned Amphoteros once more to take command of the fleet in the Aegean. Having a superior fleet of his own now, he could provide major assistance. The original fleet in Aegean waters was 160 strong, and presumably most of that was still in being, or at least could be called out again. Amphoteros returned to Greece with an extra hundred Phoenician ships to reinforce the fleet already there.[21] These Phoenician ships were therefore sent back to where they had already been, but now on the other side. They were to be joined with the Macedonian fleet which had shied away from meeting it in battle four years before, and they were to campaign against the mercenaries who had earlier fought as their allies.

Amphoteros clearly had a delicate task. He had to command two essentially separate and possibly antagonistic fleets, but he had also to fight a land campaign in Crete, while watching carefully the actions of King Agis in the Peloponnese, all coordinated as far as possible with Antipater in Macedon, who was putting down the Thracian problem. This fleet – 160 Macedonian and a hundred ex-Persian vessels – was no doubt sufficient to cope with the Persian ships in Crete, which comprised the small remnant of Pharnabazos' fleet and those which the mercenaries had arrived in. But anyone at sea in the Aegean was always conscious that the real naval power in the area was that of Athens, whose enormous fleet outnumbered all the others put together. If Athens came out against Alexander, the whole of Greece and the Aegean was vulnerable to Alexander's enemies. Another of Alexander's decisions at Tyre was to release his Athenian hostages, ships and men, an appeasing gesture which was clearly intended to draw the sting of hostility at Athens.

The Persian naval forces in the Aegean, therefore, were now of little account. We hear nothing of what Amphoteros did, so it seems obvious that his overwhelming naval power simply stifled any opposition. The mercenaries moved over to the Peloponnese and were a large part of the army which Agis lead to defeat at the battle at Megalopolis in the spring of 330. Alexander was therefore able to ignore all this and continue on eastwards. By his control of the Cypriot and Phoenician fleets, and the ships he had on patrol in Egypt, he was able to defy anything Agis could threaten. Once Agis had been defeated by Antipater, the fleet was again not needed; presumably most of the ships went home.

The numbers of ships in these various fleets are always suspiciously exact and round. Twice Alexander's Aegean fleets are said to number 160 ships; at Tyre each Cypriot king is said to have brought forty ships to the siege; the Lykians and Rhodians sent ten each; the Tyrians had eighty. And the original Persian fleet is put at 400. It has already been noted that it seems very unlikely that this last number was ever reached, though it may have been aimed at –

Memnon apparently never actually had more than 300 ships, and most of these were used to take his mercenaries to Syria. The equality in number of the three Cypriot contingents may be due to Alexander's requirements, or it may be a legacy of Persian rule. The Tyrians are said to have had double that number: maybe the Persian government had set these totals as a rule of thumb for each city. Yet when we have precisely documented numbers rather than the later statements of historians, naval numbers are normally rather less round. So the records at Athens give precise bureaucratic figures, only one of which is a round number. The figures of the chroniclers and historians may therefore be taken only as approximate indications of the sizes of the fleets, not as precise countings.

For Alexander, after the fall of Tyre, it is likely that the number of ships at his command ceased to be of interest. He could, as Arrian remarked, assume that a campaign by land had successfully seen off the threat of the Persian fleet.[22] Yet this was not the case. At no time had the Persian fleet suffered the inconvenience of having no base. After Hegelochos and Amphoteros took back the Eastern Aegean islands, the remnant of the Persian fleet was able to use Cretan ports. And at no time had Alexander campaigned with the express intention of depriving the enemy fleet of such bases, despite the story of the eagle on the shore. The only possible occasion when this was his policy was at the siege of Miletos, where the Persian fleet was kept at a distance, and where its mainland base was seized by land troops. But the ships merely moved elsewhere, to Samos, to Kos, to Halikarnassos, to Crete.

On his campaign eastwards Alexander certainly took control of several ports, but in only one instance, at Side in Pamphylia, did he install a garrison. Other ports welcomed him, and he appointed satraps to govern provinces, but had the Persian fleet seized a port anywhere, even one he had already garrisoned, it is likely that neither a garrison nor a satrap could have done anything about it. One does not defeat a fleet by occupying the coast and the ports; the enemy fleet would be far too strong locally to be deterred from landing.

Alexander was, it is clear, not sea-minded. At Tyre he put heavy infantrymen on the ships, intending to convert the sea fight into a pseudo-land battle. Instead of using his ship superiority to attack the island, he used it to guard the laboriously-built mole. When his fleet was too expensive, he disbanded it rather than find a use for it. When, in India, he was compelled to turn back and head westwards, he had a fleet built in order to carry a proportion of the army by sea. This was sensible, since movement by sea was easier and quicker than by land, but the fact is that he saw the ships merely as transport vessels, not in terms of power.[23] His last planned campaign was to build a fleet for the purpose of circumnavigating Arabia – but this was merely transport again.[24] It is said that he aimed to build a fleet of a thousand ships in the Mediterranean.[25] The very number indicates that the aim was not sea-power, but an exercise in megalomania. There was no need for a fleet of such a size, nor could Alexander man it from his resources. Even if only triremes were built, he would need 200,000 men as crews, and even more if he aimed to build quinqueremes; and more still if he installed extra fighting men, as he had at Tyre.

Alexander's rather inept conception of sea-power nevertheless gave impetus to a change which had been slowly under way for several decades. At the siege of Tyre triremes were the most numerous vessels in all the fleets, but quadriremes and quinqueremes were increasingly present. Each of the Cypriot kings had a quinquereme as a 'royal ship', no doubt for prestige reasons, but also as a command centre. These bigger ships were easy to distinguish in the heat of conflict, and their importance is made clear by the Tyrian attack in the northern harbour during the siege, when it was these Cypriot royal ships which were singled out as the first to be attacked.

The Tyrian attacking force also included both quinqueremes and quadriremes, three of each. There were also quinqueremes among the Phoenician ships in the south harbour fleet which Alexander took to sea to combat the Tyrian attack. There were not many, but it is clear that a quinquereme was needed to face another quinquereme. If the Cypriot kings had one each, probably the Phoenician kings also had one each, at least. Tyre seems to have had a bigger fleet than anyone else, perhaps as a result of Persian favour. The city had been conspicuously loyal at the time of the Sidonian revolt in 362, which may explain its possession of more ships, and it may explain Tyrian loyalty to Persia in the face of Alexander's attack. So in the whole of the eastern Mediterranean there were probably no more than ten or a dozen of these bigger ships. In the Aegean there were as yet none, so far as we know. There may have been a few others in Sicily and North Africa, but the naval competition there had died down with the demise of the Dionysiac monarchy in Syracuse in the 340s, and even Dionysios I had probably had no more than one or two of the greater ships at any one time. His son, Dionysios II is said to have built a 'six', but he had lost power in an internal revolution, and the ship was never used. It is, however, perhaps characteristic of Alexander that he is said to have planned not merely a fleet of a thousand ships, but that they were to be even bigger than quinqueremes. Or, more likely, it may be that this is what later writers thought he would have intended.

When Alexander died, in June 323, his final plans had scarcely passed beyond the stage of preparation, and the naval plans for the Mediterranean had not even got so far. Hence, of course, the doubts as to his real intentions. Even so, he commanded a naval force which, for the first time for a couple of generations, was of a size which could challenge that of Athens. And, since Athens had been in a condition of simmering hostility ever since its defeat at the Battle of Chaironeia fifteen years before, it was highly likely that the implicit naval challenge would be accepted.

Chapter 2

The End of Athens' Sea Power

For a century and a half the dominant naval power in the Mediterranean had been Athens, though it always had competitors. Defeats of the city by sea were occasioned by a combination of its enemies, such as Persia and Sparta after 411. Recovery from that defeat was relatively straightforward, for two such unlikely allies as Sparta and the Great King could not remain friends for long. As soon as their alliance was broken, Athens could rebuild – both its walls and its fleet. The city's defeat by Philip II in the 330s, on land, had scarcely affected the fleet at all. The ships remained numerous and up-to-date, and when Alexander died in Babylon in June 323, Athens had a bigger fleet than anyone else.

We know of this fleet in some detail because Athens' bureaucracy recorded the details in stone so that the people, in whose name the bureaucrats worked, could see that their instructions were being obeyed – accountability, bureaucracy, and democracy combined. The permanence of the record is our good fortune. In 330/329 there were 392 triremes on the city's books, plus eighteen quadriremes. By the time of Alexander's death the number of quadriremes had risen to fifty, while the triremes, which had risen to 360 in 325/4, had sunk to 315.[1] At about the same time a decree was passed by the assembly that the future building program should be forty triremes and 200 quadriremes. This has occasioned some dispute among modern historians, in that it has been assumed that it should be the other way around, for nowhere near 200 quadriremes[2] were ever built by Athens. It is now argued that this was the *planned* work; given the clear concentration on producing quadriremes in the years before 324, this seems a reasonable interpretation.[3] Athens not only had the biggest fleet, the city was also up-to-date with the technical developments.

This large force of ships was accommodated in ship-sheds at Peiraios, a building in which the ships could be drawn out of the water, dried out, and serviced and maintained. The wooden ships of the time had a life of perhaps twenty to thirty years, which could be extended by careful maintenance. The Athenian ship-sheds were reconstructed during the generation before Alexander, and work on them was interrupted by the Chaironaia war emergency in 338. There was also a famous hanging-gear store designed by the architect Philo.[4] This institution – sheds and store – meant that most of Athens' ships were kept on dry land, but were available to be launched and commissioned at instant notice. The naval records regularly note how many ships were at sea – in 325/324, for example, 32 triremes and seven quadriremes were out.

This was all very impressive, though the decree clearly envisaged a notable reduction in the total number of ships in Athens' fleet, from over 400 in the 330 list to 240 in the future. This decree seems to imply that many of the 410 ships of 330/329 were unusable, perhaps kept on the books only by bureaucratic inertia. Athens had still to come to terms with the financial and personnel implications of such a huge naval force. Quite simply, Athens had too many ships for her population to man. The normal complement of a trireme was 170 men.[5] To man all 395 triremes which the city had in 330 would have taken over 66,000 men, a total quite impossible for Athens to produce, for the adult male citizen population was less than half that. This realization may be one reason for the reduction in ship numbers which followed.

Those implications struck home decisively in 322. Alexander had made direct attempts to exert control over the internal affairs of the Greek cities by demanding to be accepted as a god and requiring the return of political exiles. This had annoyed many, particularly Athens, so his death provided the impulse for an attempt to remove Macedonian power from the cities of Greece. In this Athens was crucial. Of the other major Greek states, Sparta had been crushed by Antipater in 330, and Thebes was still in ruins; Aitolia had emerged recently as a new power, but was as yet unreliable. Another 'power' was the camp of unemployed mercenaries at Cape Tainaron in Sparta, many of them disgruntled at their defeat by Alexander at Issos or by Antipater at Megalopolis.

Both Athens and Aitolia were directly threatened by the 'exiles decree'. An alliance between the two was an obvious political move, and an Athenian soldier, Leosthenes, became the link between his city and the mercenaries' camp. This new war would have to be fought by land, an assault on Antipater's power in Macedon, so Athens had to produce a major army. Yet it had also to be a sea war. Antipater had 110 warships at his disposal, and the Macedonians in Asia had even more.

Athens' navy was large, but not all of the 410 ships in the inventory were fully seaworthy. The life of a trireme, so long as it survived any fighting, was reckoned at two or three decades.[6] To maintain a navy meant continually replacing old and rotten ships by newly built ones. The plan set out in the decree which envisaged building 40 triremes and 200 quadriremes was presumably aimed at this very fact: as the new ships were built, the oldest ships would be removed – or the other way about, perhaps. So, not only was a considerable proportion of the fleet probably unusable, not even those ships which were seaworthy could be fully manned. There is some confusion over numbers, but it is clear that the highest number noted in use in this new war is 170, though this is probably not the original total of ships launched. The basic reason for the disparity between ambition and achievement is that Athens was having to field a full land army at the same time; as a result the fleet had to be smaller than perhaps had been intended.

This fleet was not out of practice. Parts of it had been used several times during Alexander's time: in 332/331 a hundred triremes were launched to go to Tenedos to release the grain fleet, though it may be that the mere threat was enough.[7] The whole of Greece suffered famine between 331 and about 325, so the safety of the grain fleet was crucial. Also, perhaps as a result of the famine, Athens sent out a colony to somewhere in the Adriatic expressly to guard against attacks by Etruscan pirates; the overall purpose was to safeguard food supplies from that region. A number of ships was allocated to this service, of which three triremes, two transports, and four triaconters are known about; some quadriremes and other triremes were also used in addition.[8] These ships will have stayed at the colony, and were perhaps exchanged for a new squadron regularly. Its very purpose was to thwart the activities of the Etruscan pirates so it will have been kept up to strength. It is another of the reasons the full Athenian fleet was not available next year.

In addition to this activity, it seems clear that the fleet, or part of it, was used in 323 to lend extra persuasiveness to the busy Athenian diplomacy which produced the Greek alliance. The town of Styka in Euboia is said by the geographer Strabo to have been destroyed by the Athenian general Phaidros in the 'Malian War'.[9] Phaidros is a known figure of the period, but the 'Malian War' is otherwise unknown – hence the generally accepted emendation to 'Lamian War', which is the usual name for this new Greco-Macedonian fight. But Lamia is the main city of Malis, so the name 'Malian' is not really misleading.

The destruction of a town by an Athenian general in the context of the Lamian War suggests that forceful methods were being used to bring reluctant participants into the alliance. Strabo makes the point that Styka in his day (the reign of Augustus) was part of the city of Eretria; possibly this had been a bribe to bring in the Euboians to the alliance. However, much had happened between the Lamian War and the time of Augustus so not much reliance can be placed on such a conjecture. A wider consideration would suggest that the Athenian fleet was on a voyage round the islands, rounding up support by a display of power, and as perhaps occurred at Styka, suppressing dissent and bringing in neutrals by armed persuasion. The greater islands of Lesbos, Chios and Kos might well be resentful of Macedonian control after being fought over a few years earlier; Rhodes seems to have expelled a Macedonian garrison about this time.[10] It may well be that the islands generally were part of the alliance by the end of 323, though being members of the Athenian alliance may not have seemed a great improvement on the Macedonian.

To describe the maritime aspects of this war is particularly difficult. The main account is in Diodoros, who has, it seems, compressed the sources he was using to such an extent as to render them barely comprehensible.[11] In addition several Athenian inscriptions were set up later, when it was safe to do so, recalling some memorable incidents, and some details can be culled from later sources. Combining the several elements is difficult, and it is not surprising that those who write about this war tend to concentrate on the events on land, which are far better recorded.

But the war at sea was decisive in bringing about the defeat of the Greeks. It is clear that on land they could have been victorious if they had won at sea, for it was the defeat of the Athenian navy which permitted the Macedonians to recover after their initial defeats. The crucial control-point in the maritime war was the Hellespont, since it was there that the Macedonian reinforcements which eventually reached Antipater in Thessaly would have to cross from Asia to Europe, and it was the best place to exert pressure on Athenian food supplies.

The war began in the late summer of 323. Alexander died in Babylon in June, and it will have taken some weeks for the news to reach Athens, for the necessary diplomacy to take place, and for the Greeks who joined the alliance to mobilize their forces. The early battles took place in Thessaly (Macedonian territory), and by the autumn Antipater had been pushed into the city of Lamia and was under siege, though Leosthenes had been killed. Until this point the fleet had not been needed, though it was known that Antipater had 110 ships, either with him or at his command.

It has been theorized that Antipater's ships were in the Malian Gulf, but it is difficult to accept this, given that once he was locked up in Lamia, his ships would be extremely vulnerable to land attack – the Greek forces controlled all the shores of the Gulf. Diodoros also notes that later the Macedonians destroyed some Athenian ships at the Echinades Islands. These are off the coast of Akarnania to the west, and it is generally disbelieved that such an event could have occurred there in this war. So by a sleight-of-hand this event is usually relocated to the Malian Gulf, where there happens to be a place called Echinos (but no islands of that name).[12] The problem, of course, is that here we are faced with a whole series of emendations to the sources to compel them to conform to a theory; this is the wrong procedure.

If Diodoros was correct in his reference to the Echinades Islands, and so is referring to an event among those islands in the west, then we may assume that the fight was between a Macedonian squadron and the Athenian ships which had been sent earlier to assist the colony in the Adriatic. The Macedonian naval commander was Kleitos; he was probably not in the west himself, but he was certainly in overall command of all the Macedonian naval forces. Given that Diodoros has compressed his source material, it will not do to insist that Kleitos could not have been in command at the islands and therefore that the fight did not take place in the west. We must accept that a sea fight, in which the Macedonian ships were victorious, took place at the Echinades Islands in the west. This conclusion, of course, means that there is no reason to locate Antipater's 110 ships in the Malian Gulf.

The Athenian fleet, having been active during the later months of 323, will have been largely laid up during the winter; so were the Macedonian ships, for the Aegean is dangerous in the winter months. Meanwhile Antipater was besieged in Lamia throughout the winter, and Macedonian reinforcements did not reach him until the spring. Antipater contacted Leonnatos, the satrap of Hellespontine Phrygia (on the Asian side of the Straits), who eventually arrived in the spring, but Antipater was penned up in Lamia all through the winter of 323/322.

The Athenian fleet's objective was therefore the Hellespont, in order to block the transfer of Macedonian forces from Asia to Europe. And it was there that a battle was fought, near Abydos. The Athenians were defeated. The evidence comes from two inscriptions which honour men who assisted Athenian sailors who survived 'the battle in the Hellespont' and managed to get home.[13] The questions are, therefore, when and against whom was this battle fought? But first it is necessary to discuss the other sea battle of this war.

This second battle (or third, if that at the Echinades Islands is included) was fought near the island of Amorgos, at the southeastern end of the Kyklades Islands. It is better recorded, because it is noted in the inscription called the 'Marmor Parium' as 'the sea battle fought and won by the Macedonians at Amorgos'.[14] However, since the two conflicts at Abydos and Amorgos are recorded in different ways and in different places there is no indication which came first. Amorgos has been dated to June or July 322 by a complex calendrical argument, whereas, since the Hellespont was clear for Macedonian troops to cross from Asia to Europe long before that, the Abydos fight must have already happened. Leonnatos crossed with some troops, then spent some time in Macedon recruiting more, and finally marched south to assist Antipater, reaching him in the spring. It follows that the passage across the Hellespont was available to him considerably earlier than that. The battle off Abydos was, it may be assumed, part of a Macedonian naval campaign to clear the way for Leonnatos' forces to get across.

This, in fact, as will be appreciated, pushes the Abydos battle back well before the spring of 322, in fact perhaps even into the preceding autumn. And then there is the problem of which Macedonian ships the Athenians were fighting at that time. The eventual Macedonian commander by sea was Kleitos. He had been with Krateros in Kilikia during 323. He eventually fought off Amorgos (in June or July) with a fleet of 240 ships. Even if he had Antipater's 110 ships as part of his force, he had still had to bring at least 130 ships with him from the east, presumably collected from the Phoenician cities of Syria – that is, they were Alexander's and Dareios' former fleet.

This must have taken some time to arrange. The Greek rebellion developed during the late summer of 323, and Antipater contacted Krateros about the same time, so it is presumably when the two commanders had formulated their alliance (Krateros married Phila, Antipater's daughter) that Kleitos was commissioned as admiral and began to collect his fleet. This force therefore could well have arrived in the Aegean during the autumn of 323, while the Athenian fleet was on its diplomatic campaign in the islands. Since it was clear to all the Macedonians involved that the Hellespont was the crucial point, Kleitos will have sailed for it directly. Therefore when the Athenians arrived, either in the autumn of 323 or the spring of 322, Kleitos was already in occupation of the crossing point, and it was this which allowed Leonnatos to cross over without trouble. By this time also, Krateros was on his way west with still more Macedonian forces, since it was clear that the situation in Greece was critical. So the battle in the Hellespont was probably fought to keep the crossing open for Krateros' forces, but it had already been open for Leonnatos'.

Since the size of the Athenian fleet was public knowledge – though the number of ships they would commission was unclear – Kleitos will have needed as many ships as he could gather. Antipater, besieged in Lamia and short of food, clearly did not have access to the 110 ships he is said to have brought south from Macedon in the autumn of 323, and having removed the supposed Athenian victory in the Malian Gulf, we may therefore assume that those Macedonian ships had been sent across to the Hellespont as soon as their mission in Thessaly, whatever it was, had been accomplished, or as soon as it became clear that they were liable to be destroyed if they stayed in the Malian Gulf. Therefore we may assume that Kleitos in the Hellespont had command both of Antipater's 110 ships and the ships he had brought with him from the east. It follows that the Athenians, who will have known of the Macedonian occupation of the Hellespont, had to put forward their maximum naval strength in order to attempt to force a passage through the Straits. One effect of all this was to leave Athens short of food.[15]

It has been argued that the shipyard inventories imply that Athens sent out 184 triremes and 49 quadriremes.[16] The calculations depend on restoring figures in a damaged inscription, but they seem reasonable. (This would require at least 30,000 men, assuming under-manning and smaller crews on quadriremes and triaconters than in triremes.) But the total of 233 ships at sea has to include those in western waters, the vessels eventually destroyed at the Echinades Islands. Those ships numbered at least three triremes and four triaconters, plus an indefinite number of other triremes and quadriremes; we may assume a western fleet of at least ten ships, therefore, and perhaps up to twenty. This would reduce the Athenian forces at sea available to meet Kleitos to about 220 ships or so, a smaller force than Kleitos later had at Amorgos. And at Amorgos Athens is said to have had only 170 ships against Kleitos' 240. The implication is that forty or fifty Athenian ships were lost at the Hellespont.

The Amorgos battle was later reckoned to be the important one, yet it was, it seems, not much of a fight. Anecdotes recorded by Plutarch suggest that not a great deal happened. Kleitos' victory seems to have involved 'capsizing three or four Greek ships'. And the Athenian wrecks – that is, damaged ships which were still afloat – were towed back to Athens.[17] Since this is a journey of a hundred sea miles, it is clear that the Macedonians were not following up their victory with any energy. In fact, if the Athenian fleet of 170 ships was faced by a Macedonian fleet of 240, it is highly likely that the Athenians would have carefully avoided battle if possible. Kleitos will have understood that it was, above all, essential that the Macedonian army in Thessaly be assisted, and his priority was to prevent the Athenian fleet getting to the Hellespont. Amorgos is an odd place for a battle in this war; perhaps the two fleets had been campaigning to deter each other from getting to Thessaly, and ended up far to the south (rather like the long sparring between Eumenes and Antigonos a few years later which brought them eventually to battles on the Iranian plateau).

Amorgos therefore was decisive for two reasons. The Athenians refused an all-out battle and limped home, having lost some more ships, both 'capsized' and damaged, and with their reputation and confidence damaged beyond repair. The Macedonians, those landlubbers, therefore won the sea campaign and were able to reach Thessaly first, and the final land battle at Krannon was won very easily by the Macedonians. Just as at Amorgos, the Greeks gave in extremely easily; this was presumably one of the effects also of the defeats at Amorgos and Abydos.

The result of the fighting in Thessaly was a land victory for Antipater to complement Kleitos' sea victory. The political result was a Macedonian occupation of Athens, which eventually was institutionalized in the (fairly benign) tyranny of Demetrios of Phaleron. A preliminary stage in the shift from democracy to tyranny was the reduction in the number of full citizens by raising the minimum property qualification for citizenship, a measure insisted on by Antipater. Not surprisingly the new administration favoured the wealthy, and amongst other things it abolished liturgies, one of the principal means by which wealth was redistributed by the democratic regime. One liturgy which ceased was the *trierarchia*, the provision and maintenance of ships by rich men. This meant that a part of the income of poor men – wages as rowers – also ceased. Without rowers there was no point in building ships. The result was the reduction of the Athenian navy to about thirty ships by the end of Demetrios' regime in 307. This must have involved the deliberate destruction of some of the ships as well as the phasing out of the oldest vessels. This change was a welcome development for Athens' enemies and victims, particularly Macedon, so much so that it is as certain as it can be without precise evidence that the destruction of Athenian naval power was deliberate Macedonian policy.

The elimination of the Athenian navy from the Aegean may well also have been welcomed by many of the islanders, whose attitude to the 'Greek liberty', which Athens claimed to have been fighting to preserve or restore, was rather different from that of the Athenians. The defeats of Abydos and Amorgos and the Echinades Islands may have been the agents of this change but it was, on a larger view, one which was inherent in Athenian internal politics and social relations from the very start. The wider effect was to hand the control of the seas of the eastern Mediterranean to the Macedonians, but even as this shift took place, those same Macedonians were disputing between themselves about the location and possession of power.

Chapter 3

Antigonos Takes to the Sea

The defeat the Greek insurgency, the Lamian War, was soon followed by disputes between the many ambitious Macedonian commanders. These disputes escalated steadily to open warfare, and the royal fleet was soon in use as a piece in their games. In the first crisis, Perdikkas, the imperial regent, invaded Egypt. His brother-in-law Attalos commanded the fleet, but he only brought it as far as the mouth of the Nile near Pelusion.[1] The ships apparently could not get up the river – it seems to have been the time of the flood – and Perdikkas' soldiers had to attempt to cross over without shipping. It was a disaster, and the campaign ended with the assassination of Perdikkas. His place was taken by Antipater, who was already old; he died in 319. At that point the competition became open.

Attalos took the fleet to Tyre and from there sailed to Karia in southwestern Asia Minor. He attacked Rhodes, but the small Rhodian fleet fended him off.[2] He then headed inland where he and his associates were captured by Antigonos, who had been the satrap of Phrygia in Alexander's time and had extended his power over most of inland Asia Minor. It is to be assumed that the fleet fell into Antigonos' hands as well, for he was soon able to lend thirty-five ships to Kassandros with which the latter seized control of the Peiraios port and the Mounychia fort at Athens.[3] With these he was able to dominate the city (and presumably seize control of the remaining Athenian ships which were kept at Peiraios). Another part of the fleet was under the control of Polyperchon, Antipater's designated successor as imperial regent who retained Kleitos in command. Kleitos took the fleet to attack Antigonos in the Straits (whereupon Kassander returned the thirty-five ships). The aim of Kleitos' campaign was to gain control of the Hellespont, but the fighting took place near the Bosporos; presumably Antigonos had retreated before Kleitos' greater numbers. Antigonos was defeated in the first fight near Byzantion, losing seventeen ships sunk and forty captured. But Kleitos and his men then relaxed, and were subjected to a sudden land and sea assault at dawn, which panicked the men. Kleitos lost his army, all his ships but one, and eventually his life.[4]

The disproportionate effect of the use of a relatively small sea force is evident in all this. The thirty-five ships of Antigonos enabled Kassandros to establish his power in the main fortified city in the Aegean; they were then sent back to Antigonos and played their part in Antigonos' own victory over Kleitos. Antigonos took the point and was inspired. It was just here that the historian Diodoros comments that Antigonos now determined that he aimed to 'gain command of the sea and place his control of Asia beyond dispute'.[5]

However, Antigonos' sea path to power was not easy. His naval commander, Nikanor, the man who took the thirty-five ships to Kassandros, had participated in the victory near Byzantion, and now he sailed back to Peiraios, for he had originally been Kassandros' man, and took a large part of the fleet with him. Kassandros detected in him a new arrogance and ambition and had him assassinated.[6] The fleet at Peiraios thus now became Kassandros' fleet once again, and he used it during 317 to help him gain control of Macedon. With Athens and Macedon in his hands, and an army and a fleet, he had suddenly become very powerful.

So Antigonos lost control of a large part of his victorious fleet. Meanwhile his Asian rival, Eumenes of Kardia, had moved south into Syria, into which the satrap of Egypt, Ptolemy, had already intruded. Ptolemy prudently withdrew before Eumenes' greater armed strength. With vivid memories of Perdikkas' attack on Egypt by land and sea, as Eumenes retreated he carefully moved all the naval vessels in Syrian ports to Egypt. Eumenes camped in Syria and Phoenicia and collected ships from the ports which had been out of Ptolemy's reach. It did him no good, for, as soon as Antigonos' fleet arrived, Eumenes' scratch fleet changed sides.[7] Soon afterwards Eumenes was pushed out of Syria eastwards by Antigonos; the two of them duelled in the east for over a year.

The result in terms of sea power was that part of the Macedonian fleet was now in Egypt under Ptolemy's control, a second part was in Macedon and the Aegean under Kassandros, and a third part was somewhere between the two, loyal to Antigonos. The three men were still allied against Eumenes, so there was no more conflict between them, at least for the moment. Ptolemy was able to use his fleet to develop an overlordship of Cyprus, but Kassandros scarcely seems to have made use of his at all.

There were also smaller fleets here and there in other parts of the Mediterranean. The Cypriot kings usually had ships, but with a dozen kings in the island none of them could develop much sea strength, especially in the face of Ptolemy's fleet, which soon absorbed their ships. Rhodes, which was at enmity with most Macedonian generals, had an effective small fleet, mainly of quadriremes, and had been skilful enough to deflect Attalos' attack with a more numerous fleet. Athens' fleet was now controlled by Kassandros, and the city saw the size of its fleet dwindle until it was of no more than local importance. In the west, also, the two old rival cities of Syracuse in Sicily and Carthage in North Africa had long maintained relatively modest fleets, mainly as deterrents against each other. Syracuse was going through a new political phase of upheaval, and soon an ambitious tyrant would be in control; this usually meant war against all Syracuse's neighbours, and an arms race with Carthage, but for the moment war at sea in the west was directed against pirates only.

Elsewhere, other civic fleets were of no more than minimal local importance. Massilia, Tarentum, Rome, the coastal Etruscan cities, Corcyra, and others all had a few ships, which were designed mainly to protect the coasts and merchant ships against pirates. None of them counted in the great political stakes being played for in the eastern Mediterranean.

One modern historian has pointed out that Antigonos, who was victorious over Eumenes in a final battle in Iran in 316, was the one man amongst the Macedonians who appreciated and promoted sea power; presumably he captured the idea from Diodoros' words, quoted earlier.[8] It is, however, not really an accurate characterization. Any Macedonian general could appreciate the importance of sea power when it was necessary. There is no doubt, for instance, that Ptolemy, attacked in Egypt once already by a great fleet, understood its effects and its use, which was why he had seized the Syrian ships. What was different about Antigonos was that he possessed, in 315, a fleet which was a fragment of that of Alexander, many of which ships were old by now, but also controlled long coastlines and many port cities. And he came back from the campaign in Iran enormously wealthy. For, as Athens had known, it was wealth which was the fundamental essential for developing sea power.

Antigonos, burdened by a huge army, a great treasure, and several awkward tasks to do on the way, marched slowly back from Iran to the Mediterranean coast during 315, reaching Kilikia in November (where he picked up yet another treasure). During the journey he confirmed or replaced satraps, as well as appropriating state treasuries. He thus acted as a king or a regent rather than as a subject, which he technically was. (There was a king, Alexander IV, and a regent, Polyperchon, acting for him still, though Antigonos claimed to have authority to act for the king in Asia.) One of the satraps he removed, Seleukos of Babylonia, escaped from what was probably certain death at Antigonos' hands to report what was going on to Ptolemy, who had been his colleague in the action against Perdikkas. Ptolemy, Kassandros in Macedon, and Lysimachos in Thrace then allied themselves together and presented Antigonos with an ultimatum, demanding their share of the spoils he had gathered – though this was clearly just an excuse. Confident in his power and his wealth, Antigonos scorned them. War resulted.

Antigonos moved south from Kilikia to take over Syria, much of which Ptolemy had occupied once more when Eumenes was driven out. Ptolemy withdrew again, leaving a garrison in Tyre, which was still, despite Alexander's destruction and the mole he had built, the strongest military point in the land. Ptolemy used his sea power to supply the city, and kept some of his ships there.

Antigonos ignored Tyre, but he now controlled every other port in the region, and he set about building a new fleet. Diodoros gives a good explanation of what was involved in Antigonos' work.[9] The kings of the Phoenician cities were gathered along with the governors of the parts of Syria under Antigonos' control. Food was collected. Shipyards were organized at Sidon, Byblos, and Tripolis (and, one would have expected, at Arados), and at an unnamed place in Kilikia. Wood, notably cedar, was cut in the forests of the inland mountains, and 8,000 men were employed in cutting the trees down and sawing the wood; a thousand pairs of draught animals were used to bring the sawn wood to the shipyards. Diodoros does not detail the workers in the shipyards, probably because they were already employed there, but there must have been several thousands of them. More ships were built at Rhodes on a contractual basis, Antigonos supplying the timber and recruiting the crews.

Ptolemy gave Seleukos command of a hundred ships from his fleet, and he sailed along the Phoenician coast, no doubt primarily to see what was going on. The sight of the fleet so cheekily close by is said to have depressed Antigonos' men; he replied to Seleukos' gesture by proclaiming that he was in the process of building 500 ships. But Seleukos' cruise was an interesting display in the use of sea power. He could come close to the enemy coast with impunity, and he could cruise all the way from Egypt to the Aegean without hindrance.[10]

Ptolemy had to counter Antigonid intrigues in Cyprus; his grip was enforced by 10,000 soldiers, who were used to suppress several of the city-kings. He also stationed a hundred ships of his fleet there.[11] This clearly put him in a position to intercept any squadron of Antigonos' coming out of the Syrian ports. Ptolemy's Cypriot squadron thus separated the several parts of Antigonos' fleet, and presumably compelled him to rely on slower land communications. Antigonos threatened Kassandros' power in the Peloponnese by recruiting troops at the mercenaries' camp at Cape Tainaron in Sparta – all it took was money and a recruiting agent. Seleukos from Cyprus sent fifty of Ptolemy's ships, under Polykleitos, to exert a counter influence and support Kassandros' position, and Ptolemy sent Seleukos himself into the Aegean with part of the fleet, though he seems to have accomplished nothing militarily. His presence was as much aimed at keeping the alliance together as it was intended to achieve a victory. He was besieging the city of Erythrai in Asia Minor when an enemy force approached; capturing the city would have had little effect; he retired.[12]

Antigonos did not wait for his full 500 ships to be ready – though that figure was probably merely boastfulness on his part anyway. He finally, and efficiently, organized the conquest of Tyre, where he captured some of Ptolemy's ships, then collected ships from all around his coasts, forty from the Hellespont and eighty from Rhodes and the Hellespont. These were presumably the remnants of his original fleet, plus the new building done at Rhodes, and perhaps others from shipyards in Asia Minor. His efforts in Phoenicia had produced 120 ships, giving him a total of 240. Fifty of these he sent to the Peloponnese to counter the similar fleet of Ptolemy's ships which were already there.[13] In theory, therefore, in eastern waters Antigonos had a preponderance of 190 ships to Ptolemy's 150. But in no time the picture was spoilt.

The Ptolemaic ships in the Peloponnese had been doing little good, and were brought back to the east, perhaps to bolster the 150 in Cyprus. On the way, off the coast of Kilikia, Polykleitos intercepted and defeated a force of Antigonos' Rhodian ships, though the size of the defeated force is not known. The action was cunning. Hiding the ships behind a headland, Polykleitos attacked the army which was marching along the coast road in company with the ships, and defeated it. The Antigonid ships came in to assist their colleagues, whereupon Polykleitos brought his own ships round to attack them from the sea. The Antigonid ships were all destroyed or captured.[14] The naval balance at once tilted back towards Ptolemy.

The notice of the gathering of Antigonos' fleet gives details of the types of ships he had, though it is not clear which were new, Rhodian, or old. The majority were triremes and quadriremes, 127 and 90 respectively out of the 240. There were 10 quinqueremes, 3 'nines' and 10 'tens'. Of the triremes thirty are described as *aphract*, 'undecked'; the rest (not actually listed) are therefore assumed to be covered, that is *cataphract*.[15] This is a curiously mixed force, and only the nines and tens are new. These were ships which had the usual three banks of oars, but the oars were much larger and were rowed by nine or ten oarsmen. Clearly the assumption was that these huge vessels would succeed against smaller ones by sheer weight, but they must have been slow to move and very difficult to manoeuvre. Smaller, lighter vessels could therefore get out of their way without too much difficulty. However, like the quinqueremes of the Phoenician and Cypriot kings earlier, it may be that they were intended as massive command platforms and headquarters vessels, leading squadrons of the smaller ships.

The rival fleets were thus too evenly balanced for either contender to prevail. They had to be used with care and generally in support of other operations; neither general cared to pitch his whole fleet into a battle against an equal, at least for the present. In 314, for example, despite having lost his Rhodian ships in Kilikia, Antigonos brought his full Phoenician fleet, 150 ships under the command of Medeios, to assist in his campaign against a recalcitrant satrap, Asandros of Karia, leaving Ptolemy's fleet unchallenged in Levantine waters. A three-pronged attack by land commanded by Antigonos from inland Asia Minor was matched by fleet operations along the coast, where the cities of Miletos, Iasos, and Kaunos were captured from the sea. On the way to this campaign Medeios met and captured a fleet of 36 of Kassandros' ships, which had been sent to assist Asandros.[16]

Antigonos used detachments of his fleet to distract his European opponents during his major campaigns. Several cities on the west coast of the Black Sea had rebelled against Lysimachos; Antigonos sent Lykon with a fleet through the

Straits to assist Kallatis, the last resisting city. An army was also sent, commanded by Pausanias, possibly in the ships. Diplomatically, a Thracian king, Seuthes, was also contacted and persuaded to fight against Lysimachos. None of these expeditions was wholly successful – Lysimachos took Kallatis and defeated Seuthes; Pausanias' army was captured, and the fleet had to withdraw – but this involved only a fairly limited expenditure of men and ships, and it all kept Lysimachos occupied for a full season.[17]

Antigonos also threw a series of punches at Kassandros. The fleet of fifty ships he had sent to the Peloponnese earlier was revived and placed under the command of Telesphoros, probably one of Antigonos' nephews. He made a little progress, but could not interfere with Kassandros' campaign in central Greece. So Antigonos sent a large part of his main fleet, a hundred ships under Medeios, to relieve the pressure which Kassandros was exerting on places in Euboia. Twenty of Antigonos' ships, commanded by Telesphoros, came up from the Peloponnese and attacked Kassandros in Oreus, where some ships were burnt. But the Antigonid force relaxed, and then they were attacked by a fleet of Kassandros' out of Athens. It seems that Medeios' main fleet had returned to Asia, so it was only the smaller squadron from the Peloponnese which, strongly outnumbered, was now defeated, losing four ships.[18]

Medeios was now sent off again, to challenge Kassandros directly in central Greece. He had the full fleet of 150 ships and 5,500 troops commanded by Ptolemaios, another of Antigonos' nephews, but this annoyed Telesphoros, who resented his apparent demotion. He defected, sold his ships, and tried to set up as a tyrant in Elis; Polemaios slapped him down, but at the cost of a distraction from his main campaign. Medeios' fleet was then used by Antigonos to attempt to gain control of the Straits, but the key city, Byzantion, decided to remain neutral, which actually benefited Lysimachos, and Antigonos withdrew his fleet.[19]

Antigonos was making good use of his sea power, but his human instruments were not reliable. By scattering his shots in several directions he was succeeding in few. He did tend to distract his enemies, and the major campaign in central Greece under Ptolemaios was fairly successful. He secured control of much of central Greece, from Boiotia to Phokis, but then Telesphoros' defection stopped his progress. So Antigonos' use of his fleet in all this was intelligent, for the ships could move much more speedily than any land-based forces, and so could appear at strategic points unexpectedly: he sent squadrons into the Black Sea and the Peloponnese; the whole fleet delivered an army to central Greece and then sailed through the Straits. This was a sensible use of sea power and it looked for some time as though it might help deliver the lands controlled by Kassandros and Lysimachos into Antigonos' control. He gained a powerful propaganda advantage by pointedly leaving the captured Greek cities without garrisons, a policy which persuaded Rhodes to give him more naval assistance.

Ptolemy replied to Antigonos' success in Europe by an assault on Antigonos' forces in Syria. In a battle at Gaza in 312 he and Seleukos destroyed the Antigonid army in Syria (which was commanded by Antigonos' son Demetrios and a committee of former soldiers of Alexander). In the aftermath Ptolemy moved north as far as Phoenicia, but cautiously; meanwhile Seleukos set off on a wild and clandestine ride to recover his old satrapy of Babylonia, accompanied by only a thousand followers. Demetrios recovered his balance in North Syria, and, reinforced by troops sent by his father, drove Ptolemy back into Egypt; Seleukos, on the other hand, was spectacularly successful, recovering Babylonia, recruiting large numbers of Antigonos' disaffected soldiers and gaining control of Iran into the bargain. Antigonos thus now faced fighting on a new and major front.

Everyone was weary of the war, and a standstill peace was arranged in 311 – except between Antigonos and Seleukos, who fought on for another three or four years. But the peace in the Mediterranean lasted no more than a year. By 310 Ptolemy was raiding the coasts of North Syria and Kilikia from his base in Cyprus, probably hoping to assist Seleukos, and next year he took his fleet on a cruise along the south coast of Asia Minor, seizing control of a series of strategic ports – several places in Kilikia, Korakesion and Phaselis in Pamphylia, Xanthos in Lykia, Iasos and Kaunos in Karia, and then established himself on the island of Kos.[20] He failed to take Halikarnassos, which Demetrios arrived to relieve, presumably by sea, and Demetrios had earlier retaken most of Ptolemy's Kilikian conquests.[21]

In Cyprus the last of the island's kings had committed suicide; Ptolemy appointed his brother Menelaos as viceroy over the whole island, the last stage in his steadily tightening hold, which was compelled in part by the continuing desire of the kings and their cities for independence, and partly by Antigonos' continuing threat to invade. Ptolemy had thereby effectively severed Antigonos' sea connections between the Aegean and Syria, but he overreached himself. He took his fleet across the Aegean in an attempt to gain a foothold in the Peloponnese. He took control of the island of Andros on the way, but his interference in the Peloponnese brought him into conflict with Kassandros, a dangerous matter which could open the way for Antigonos, who had long attempted to divide the allies. Ptolemy gained control of Corinth and Sikyon and held them for a time, but subsequently withdrew to Kos and then to Egypt once more. One advantage he gained in all this, however, was that he persuaded Ptolemaios to defect from Antigonos, and then had him killed. After Ptolemy withdrew, Demetrios appears to have retaken at least some of his Lykian and Karian conquests.[22]

Ptolemy's expedition had exposed one of the limits of sea power, for he was using just a fleet without a major land expedition to accompany the ships. Antigonos generally moved with both land and sea forces in combination, aiming to conquer large areas of territory. Ptolemy had taken, it seems, only a small land force with him, sufficient to besiege and take the coastal cities, which were usually fairly small, but with not enough men to campaign more widely. His aim seems to have been to disrupt Antigonos' campaign, and only incidentally to conquer lands, though by seizing bases in the Aegean and on the Anatolian coast he had expanded the range of his power from Egypt and Cyprus as far as the southern Aegean. In the Peloponnese he appears to have expected the local population to welcome him, after which he hoped to recruit soldiers from among them. When they proved unwilling, either because of loyalty to Kassandros, or from a wish to remain neutral, he was stymied. When Demetrios began to operate against his Lykian and Karian cities he could only go home. But he held onto Cyprus, and to Phaselis and Korakesion, so he continued to dominate the seaway between the Aegean and the east, and retained the ability to reach the Aegean, which had now emerged as the central arena of conflict in the whole region. This was a strategic reach which was to be consolidated by his son.

Ptolemy's manoeuvres did not apparently bother Antigonos very much. He now moved into Syria where he began building himself a new capital city in the north. His son Demetrios had developed into a brilliant commander, if somewhat headstrong, and the old man now effectively retired from active operations and handed to Demetrios the active role. But it was still Antigonos whose strategy was applied.

Chapter 4

The Campaign of Demetrios

Demetrios son of Antigonos is one of the more romantic and tragic figures of the Hellenistic period – he is also one of only two of the successors of Alexander to have a biography devoted to him by Plutarch. He was handsome, brilliant, headstrong, careless, and arrogant. He inspired admiration and fear in equal measure. He was too ambitious for his resources, and eventually he failed in all his endeavours, totally, but between 307 and 305 he campaigned across the eastern Mediterranean with such effect that he recalled the great days of Alexander.

How far he was instructed and guided by his father in his series of campaigns between 307 and 305 is difficult to decide. One would give much for a collection of the letters they must have exchanged. Antigonos is said to have laid down the aims of each part of Demetrios' campaign, but this may be mere assumption on the part of the historians: Diodoros and Plutarch differ on what the aim of the Athenian expedition was, which suggests they were extrapolating from what happened, not recounting what was intended.[1] Antigonos spent much of the time during the campaign at his new city in Syria, Antigoneia, while Demetrios campaigned all over the eastern Mediterranean. Communications were difficult and slow, always liable to interruption and invariably out of date when received, so Demetrios necessarily had to make his own decisions, and Antigonos could only have laid down the broad objectives. Yet there is no doubt that Demetrios always deferred to his father, so it seems best to assume that the remote guiding hand was that of Antigonos, while Demetrios was the man on the spot, conducting actual operations. The two men trusted each other completely, so perhaps we may also assume they thought alike.

At the end of 308, when Ptolemy retreated from his Aegean enterprise, he left that area contested between Demetrios and Kassandros. Antigonos had now made peace with Seleukos in the east, accepting the loss of Babylonia and Iran, and Ptolemy's return to Egypt effectively meant he had removed himself from the war in the Aegean. In terms of sea power, which in this coming set of campaigns was what counted above all, Demetrios had no rival. Kassandros was never very sea-minded, and the fleet he had acquired ten years before had not apparently been kept up to strength. However, to beat him it would be necessary for Demetrios to conduct a major campaign by land. It seems that this was not in Antigonos' plans, at least not yet.

Demetrios was provided with a fleet of 250 ships, a large army, and a treasury of 5,000 talents, according to Plutarch.[2] This list is by no means as clear a statement of resources as it is supposed to be. The treasure amount is reasonable, especially since he was about to attack the rich city of Athens, and when it was conquered he would be able to replenish his treasury by taxation. But just how large a 'large' army was is not really known. A year later Demetrios sailed to Cyprus with 15,000 infantry and 400 cavalry in a fleet of 180 or 190 ships.[3] With 250 ships in the Aegean he could well have had 20,000 soldiers. Yet there is no sign in the fighting that he disposed of so many, and his 15,000 men had been recruited up to that number since the Athenian war. So it would seem he had a good deal less than later at Cyprus. (In 306 Ptolemy brought 10,000 men to Cyprus in 140 or 150 ships[5] – but neither with him nor with Demetrios are we informed of the number of transport vessels used, since the ship numbers are only of warships). So what Demetrios brought to Athens is not at all clear, except that it was no bigger than the 15,000 men he used next year.

Demetrios' instructions, according to Diodoros, were 'to free all the cities of Greece, but first of all Athens'; according to Plutarch the aim was to capture Athens, though there is a hint that further conquests might follow. It is this vagueness which suggests that no one later on really knew what was intended. 'Freeing the Greek cities' was Antigonos' stated policy, but it is also clear that his overall aim was to reconstitute Alexander's empire. We may assume that Athens was the only specific target given to Demetrios; further exploitation after victory was thus in all likelihood left for later discussion, though he clearly had some latitude in tactical matters.

The ships he had with him, said to number 250, are not otherwise detailed. Some of them at least were transports. The earlier Antigonid fleets in the Aegean had never numbered more than 150 warships, and no total greater than 240 is ever stated for Antigonos' total fleet. One of the first acts of Antigonos after the capture of Athens was to organize the building of more ships there, so it looks as though the fleet of 250 used in 307 was composed partly of warships and partly of transports. Antigonos would scarcely have used all his fleet in this one enterprise, and there were probably other ships of his in the harbours of Asia Minor and Syria, given that Ptolemy held Cyprus and had a fleet there; at the Hellespont he and Lysimachos had been building rival cities on opposite banks for the past two years, so a small fleet at least was surely on guard there. There is no actual evidence for any of these minor forces, but we must assume that they existed. So it seems unlikely that Demetrios had more than 150 warships in his fleet, which was the number that Medeios had in 312; it is possible that he had less than that.

To conquer Athens Demetrios would not need a huge army, so the 'large' army of Diodoros was probably less than the 15,000 men he used next year at Cyprus. But he planned to use even fewer men. The city was occupied by Kassandros' Macedonian troops, but they were principally concentrated in the Peiraios and in a fort built adjacent to the port on the Mounychia Hill. The men of the city were also armed and trained, as always with Greek cities, and might be expected to fight if the city looked to be in peril.

There were, however, other factors involved. It was known that many in Athens were unhappy at the Macedonian occupation, and already several years ago these disaffected elements had been in contact with Antigonos.[5] And in the past several years Antigonos had been consistent in his policy of promoting the freedom of Greek cities, refraining from garrisoning many of those he had conquered, except by request. He did not tax them extravagantly, and generally he gained the reputation of treating the cities as independent allies, rather than as subjects. On the other hand, such cities were expected to take his side, to contribute to his strength, and to do his bidding. He had, that is, good repute among the Greeks, and this was certainly understood in Athens. So the psycho-political ground was already prepared for Antigonos' intervention.

In addition to this potential internal support, Demetrios' geopolitical flanks were to some extent also protected, even though he was about to enter his enemies' territories. To the south Ptolemy had left garrisons in Corinth and Sikyon, but they were not large enough to be able to intervene at Athens. To the north Kassandros controlled Thebes (now rebuilt) and Boiotia, but Antigonos had alliances with Karystos, Chalkis, and Eretria in Euboia, and with the Aitolians further to the northwest. From Macedon, Kassandros could probably not intervene quickly; no more could Ptolemy from Corinth.

The expedition was organized secretly, which is to say that Demetrios was able to set sail with his ships and soldiers without alerting the Athenians or Kassandros. However, it was also clearly necessary that the actual attack on Athens be sudden and surprising. If a warning was given the city would have time to organize and prepare, and it was still a formidably powerful place. Having crossed the Aegean, the main force of Demetrios' fleet was held back, out of sight of Attika. A selected force of warships moved on to Cape Sounion, the southeastern tip of Attika, and out of this group, twenty of the best ships sailed west, as though towards the island of Salamis. The Athenians who were on watch for any hostile approach assumed that the ships were Ptolemy's, heading for Corinth. In the evening, when the visibility was less, the twenty ships turned and headed swiftly into Peiraios, landed the troops they carried and seized control of the port.[6]

Presumably the rowers were part of the landing force, for even with them Demetrios cannot have had sufficient forces to hold the port-city. If the ships were triremes, each would have 200 to 230 men, of whom 170 were crewmen;

twenty ships were therefore manned by between 4,000 to 4,600 men. Some were left on the ships in case they had to make a quick getaway, so Demetrios will have landed with perhaps 2,000 or 2,500 men, but only a few hundred of them were fully equipped hoplites.

The Athenians quickly mobilized; the Macedonian garrison in the Mounychia fort resisted. The initial success of the landing in seizing the port was transmitted to the rest of Demetrios' ships. The squadron left at Cape Sounion, presumably warships carrying all possible soldiers, will have arrived first; the rest, left out of sight of Attika, will have taken longer. The first contingent of reinforcements will have reached the port next morning. Demetrios proclaimed that he had come to liberate the city, and this, according to Plutarch, together with the increased number of troops in his force, caused many of the Athenians to cease opposing him. The Macedonian governor Demetrios of Phaleron escaped from the city he could no longer control, going for refuge to Thebes. It took longer to clear the city of the Macedonian garrison. The landing took place in July, but it was not until August that the Mounychia fort fell, after being battered by siege machines. It was then dismantled, to the joy of the Athenians, for whom it had been a symbol of their subjection. The Athenians heaped honours on Demetrios.[7]

It was a notable and justly renowned *coup de main*, but it did not seriously damage Kassandros, who was Antigonos' real enemy. Demetrios also captured Megara (from Ptolemy's forces) after a short siege, during the blockade and bombardment of Mounychia.[8] But he left Ptolemy's forces in Corinth and Sikyon alone; nor did he move northwards, where Kassandros still controlled Thebes. All this might suggest that the capture of Athens had been the only objective of the expedition. Perhaps it had happened so quickly that no decision on exploitation had yet been made – it would take perhaps a month for the news to reach Antigonos and for his reply to get back to Demetrios.

Antigonos apparently did not consider mounting a direct attack on Kassandros in Macedon. For Antigonos was in fact trapped by his many surrounding enemies. If he moved against Kassandros and Lysimachos in Europe he could be certain that Ptolemy would emerge from Egypt to spoil the party. Ptolemy's sea power stretched as far as southern Asia Minor by way of Cyprus. Therefore to be able to deal with Kassandros in Europe, it was necessary first to subdue Ptolemy, or at least Cyprus.

Now that he had Athens it had to be held and protected – that is, garrisoned – against any counterattack. The victory stretched Antigonid resources still further. He sent food to the city, and timber, so that the shipyards could build him, so it is said, a hundred ships; those we know of were all quadriremes, though whether the full hundred were ever built is not clear.[9] This, of course, provided employment in the Peiraios for the shipyard men, who had suffered by the reduction in the Athenian navy in the past decade. It was a cunning move, gaining both popularity and ships.

Demetrios enjoyed the adulation and comforts of Athens during the winter of 307/306. Meanwhile Antigonos decided that the next target should be Cyprus.[10] The reason Cyprus was particularly important in all this lies in its geographical position with respect to Syria and the sea currents and prevailing winds in that part of the Mediterranean. The circulation of the water in the eastern part of the sea forms a current along the North African coast eastwards, then north along the Syrian coast, and finally westwards along the coast of southern Asia Minor. This therefore favoured voyaging from Egypt northwards. The prevailing winds, on the other hand, cut across this pattern, blowing generally from the northwest towards Syria and Egypt, thus favouring travel from Greece towards Syria and Egypt. In strategic terms the main point is that warships were galleys, and their range of action to a large extent depended on the stamina of their oarsmen, and this depended, above all in the Mediterranean summer, on supplies of water.[11]

On coastal voyages north from Egypt, assisted by the prevailing current, which provided two or three knots' help, but to some extent against the prevailing wind, the range of a galley was about 550 kilometres; by the open sea route, direct from Alexandria to Cyprus it was somewhat less, perhaps 400 kilometres, unassisted by the current. (These figures are for medieval galleys, which were much the same size and capability as those of the Hellenistic and Roman periods, though perhaps a little more efficient – both employed free oarsmen, of course, not slaves.) These ranges brought Cyprus within reach of a galley sailing from Egypt, in a voyage of ten or eleven days, at least by the coastal route. Possession of both termini thus allowed Ptolemy to dominate the whole sea, and its coasts.

The problem was not, however, the time involved, or the route to be taken. The main difficulty was the need to refresh the crews. War-galleys were full of men, and had very little space for supplies, above all for water, so they needed to be resupplied whenever possible. Under ideal circumstances galleys sailing from Alexandria to Cyprus could find water at any of the ports along the Palestinian and Phoenician coasts. In wartime, with those coasts under Antigonid control, the strain on the crew was very great, and the margin of success narrow. So the control of Cyprus was a necessary base for the Ptolemaic ships. It gave a sure welcome to galleys moving out of Egypt, even if the coast they sailed beside was hostile. It also provided a base from which to raid that coast, and from which Ptolemaic fleets could intercept Antigonid ships sailing between the Aegean and Syria – as Polykleitos had.

Further, that Antigonid sea-route lay along the southern coast of Anatolia. Even ignoring Cyprus, this had become a dangerous route since Ptolemy had seized control of several cities along that coast. The Antigonid ships moving from Greece to Syria (from Demetrios to Antigonos) were sailing against the current, though they may well have had some help from the wind. They moved fairly slowly, and several of their possible refreshing posts were under hostile control, quite apart from the threat from the Ptolemaic fleet based in Cyprus.

The key to maritime control of the eastern Mediterranean was therefore Cyprus. From the Cyprus base the coasts of Syria and Asia Minor were within easy reach. For Ptolemy the control of Cyprus was also a vital defensive matter, since if it was in enemy hands his sea power would be confined to little more than the waters of Egypt and the North African coast to the west; aggressively Cyprus was also vital to Ptolemy, since it was the obvious base for any expedition by him against any of the lands of the north. Possession of the island allowed Ptolemy to dominate the whole sea, and its coasts.

All this was in addition to what is often claimed to have been the main purpose of Ptolemy in seizing control of Cyprus – that it was a source of wood, in which Egypt was deficient, with which to build his ships. (It was also a major source of metals, particularly copper.) This may well have been a factor, though it is noticeable that he campaigned to control Cyprus even while he held Syria, which was an even better timber source, as Antigonos showed in 315/314 when he had built his fleet in only a year. And there are plenty of other sources of timber in the Mediterranean area. On balance the strategic need for control of the island seems more compelling.

Ptolemy had held the island in an increasingly determined manner since Antigonos returned from the east, and his brother Menelaos now commanded there for him. Antigonos had more than once attempted to subvert Ptolemy's control – which had been the main reason that control had tightened. Several of the Ptolemaic posts in Karia and Lykia seized in 309/308 had been retaken, or taken, by Demetrios; but with the extension of Antigonid power into mainland Greece by the capture of Athens, the threat from Cyprus clearly became too great to be endured. At war with Kassandros, Lysimachos and Ptolemy all at once, Antigonos clearly identified Cyprus as the next place he needed to control. Apart from the island's inherent value, its possession would confine Ptolemy to Egypt and make it all the easier to make war in Greece and the Aegean.

The Antigonid assault on Cyprus which followed the capture of Athens therefore made good strategic sense. Possession of the island would cut Ptolemaic naval communications with the Aegean (and deprive Ptolemy of those timber and metal resources, of course). Ptolemy's possession of such places as Korakesion and Phaselis, Corinth and Sikyon, would then become largely irrelevant in the wider conflict. If necessary they could be picked off at leisure, or even just left, their importance reduced to nullity.

Demetrios sailed from Athens in the spring of 306.[12] (Soon after he left the city Kassandros returned, mounting attacks into Attika). Demetrios went first to Karia whence he sent an embassy to Rhodes asking for the same sort of support his father had received from the island more than once in the past ten years. This time the Rhodians refused, declaring their neutrality; they were trapped between their need for commercial access to Egypt, their main market, and their obligations to Antigonos. Demetrios sailed along to Kilikia, where he recruited more sailors and soldiers and collected more ships. He crossed to Cyprus with 15,000 infantry, 400 cavalry, 110 triremes, and seventy larger warships, plus the transports for the soldiers, their horses, and the siege machines.[13]

Much discussion has taken place over exactly how many ships Demetrios had, for Diodoros' wording is not clear; furthermore Plutarch and Polyainos also give differing totals, 190 and 170 ships respectively.[14] The consensus, however, is that he had 110 triremes and perhaps seventy larger ships, some of which are detailed later – 7 sevens, 40 quadriremes, 10 sixes, and 10 quinqueremes, with a few more unaccounted for. It was certainly a more formidable force than that of Ptolemy, who had more ships (190 or 200), but they were separated into two fleets at the crucial moment.

The Antigonid fleet was thus composed mainly of triremes, though more than a third of the ships were now of the heavier varieties. Two groups of ships in particular are specified by origin: thirty quadriremes were from Athens, and the sevens were from Phoenicia. These last were presumably collected in Kilikia; the Athenian quadriremes were no doubt the fruit of the work in the shipyards using the timber sent by Antigonos the previous autumn. So the rest of the ships were either brought from Athens, and were therefore part of Demetrios' original force, or were collected in Kilikia along with the sevens. It is impossible to decide on exact numbers, but it seems likely that the sixes and perhaps the quinqueremes were collected in Kilikia, and that most of the triremes and the other ten quadriremes were from the original force. This will therefore imply that Demetrios had only that number of ships the year before, plus whatever new construction had been or was still being done there, plus any Athenian vessels he had found in the shipyards. Given that it was known that Ptolemy had 200 or so ships, it is evident that Demetrios was taking on a very difficult task in attacking Cyprus. He certainly needed the reinforcements he collected in Kilikia.

Discussions of numbers of ships is a fairly abstruse exercise, and too much reliance cannot be placed on the numbers quoted in the ancient sources, none of which were anywhere near contemporary with the events they describe. Nevertheless it does look as though Antigonos' vaunted sea strength was partly illusory. The boast in 315/4 that he was building 500 ships was never fulfilled, and he managed to collect only 240 a year later, a substantial part being older ships. Some of these had to be stationed at important strategic points and cities – the Hellespont, for one, the Ionian cities for another, and now Athens – and so it was difficult to concentrate great sea strength at any one place.

A case of this had arisen in 309/308, when Ptolemy cruised from Egypt to Cyprus, and then to the Aegean, sending at least one expedition right across to Corinth. At no time was he threatened by Antigonos' fleet. This, under Medeios' command in 312, is said to have numbered 150 ships, and under Demetrios in 307 about as many. Ptolemy's fleet was probably no bigger than that. He certainly had 200 ships in 306, but he will hardly have left Egypt and Cyprus devoid of naval protection. So it is clear that Antigonos' fleet in Aegean waters had not been strong enough to challenge Ptolemy's. The 150 ships under Medeios may have been mainly warships, but some of them – up to half perhaps – were transport or supply ships. If that is so then Ptolemy had nothing to worry about.

It follows that Demetrios' fleet at Athens was not really any stronger than that of Medeios (and in fact Medeios was with him on the Athenian expedition). The only numbers quoted are the twenty ships used in the initial assault on the Peiraios, but he certainly had more warships than that with him. It is thus difficult to assume that Antigonos' disposable naval force in the Aegean was even a hundred warships. It was only Ptolemy's absence from the Aegean in 307 which allowed Demetrios to sail from Ephesos against Athens, and he was very careful to keep the destination secret, as though he feared interference. In attacking Cyprus, which was within reach of Egypt and of Ptolemy's main naval force, Antigonos and Demetrios were therefore essaying a major gamble.

From Kilikia Demetrios sailed across to land on the Cypriot panhandle, well away from the centres of Ptolemaic power in the island. There he drew his ships onshore, no doubt to clean them and dry them out, making them as efficient as possible. He captured two towns, Ourania and Karpasia, and marched his army against the centre of Ptolemaic power, the city of Salamis. Menelaos, Ptolemy's brother and the viceroy on the island, came out to meet him with an armed force roughly equal in size, 12,000 foot and 800 horse against Demetrios' less than 15,000 foot and 400 horse – Demetrios had left some of his troops to guard the ships, and others would no doubt be holding the cities he had already taken. Demetrios was victorious in the battle, and captured about 3,000 of Menelaos' soldiers, whom he attempted to recruit into his own force. They refused to serve, however, and many deserted back to Menelaos, since their 'baggage' – their lands and families, presumably – was still in Egypt. Demetrios then shipped those he still held off to Syria and settled down to a siege of Salamis. Menelaos resisted, with some success, destroying a great 'city-taker' (siege tower) Demetrios brought up against the wall.[15]

Ptolemy brought up his own fleet, landing at Paphos on the western side of the island (and so presumably sailing from Alexandria by the short open-sea route). He then sailed along to the old Phoenician city of Kition, southwest of Salamis, whence he sent soldiers to assist Menelaos, marching by land; and he directed that Menelaos' ships be sent to him. Ptolemy had 140 or 150 warships with him, which Diodoros says were partly quinqueremes and partly quadriremes, and Menelaos had 60 ships in Salamis harbour. Ptolemy also had 200 transports, which were carrying both his army and his supplies. Joining these two Ptolemaic naval forces would give Ptolemy a force approximately equal in number to Demetrios' fleet, though somewhat less in quality, given the heavier ships Demetrios also had.[16]

Demetrios brought his own fleet round from Karpasia to the waters outside Salamis harbour, reaching the area first, and so interposing himself between the fleets of Menelaos and Ptolemy. His battle-winning tactic was to block the narrow harbour entrance with just 10 of his ships, so preventing Menelaos' 60 from getting out in time to join the battle. Demetrios therefore had a fleet of about 180 ships to use in the battle, including the bigger vessels, against Ptolemy's 140 or 150. To add to his preponderance in numbers Demetrios placed all his major units, the sevens and sixes, some quinqueremes, and the Athenian quadriremes, on the left wing, keeping his weaker ships – the triremes – in the centre. His right, which was closest to the shore, consisted of the rest of his quinqueremes and some triremes. The two wings were thus stronger than the centre – one is reminded of Hannibal's dispositions at Cannae, or perhaps rather Epameinondas' at Leuktra.

Ptolemy could not be so subtle, partly because his fleet was much more uniform in composition, being only quadriremes and quinqueremes. In the fight which followed Demetrios' heavy left wing destroyed Ptolemy's right, and turned inwards to make an assault on Ptolemy's centre. Ptolemy's left was victorious against the quinqueremes and triremes of Demetrios' right, but when he saw the defeat of his own right, and the crumbling of his centre, he broke off the fight and returned with his surviving ships to Kition. In the harbour, Menelaos' ships, commanded by Menoetios, had attacked the 10 Demetrian quinqueremes at the entrance, and had eventually driven them off in defeat, but when he emerged from the harbour, apparently intending to take Demetrios' fleet in the rear – good tactics – the main fight was all over.[17]

Menoetios' action shows just how close the result was, and it was Demetrios' preliminary dispositions which had been the decisive factor. The 10 ships blocking the harbour delayed Menelaos' fleet just long enough, so that the concentration of the heavy ships on one wing enabled them to achieve a quick victory there. The fact that Ptolemy was winning on the coastal flank, and Menoetios' ships were getting out of the harbour at the same time, shows that without the big ships, and without concentrating them, Demetrios could well have been beaten. It is noticeable that he protected his triremes with some care, for if Ptolemy's quinqueremes had got amongst them – as was clearly threatened on the coastal flank – the carnage could well have been very nasty indeed.

Ptolemy had brought his transport and supply ships with him, and half of them were captured; of his warships, 40 were taken complete with crews, and 80 more were damaged and captured, some at least being repairable. Demetrios had 20 ships damaged, but repairable. On the captured transports were 8,000 of Ptolemy's soldiers. Menelaos now surrendered Salamis, his ships and his soldiers. Demetrios had therefore captured something over 16,000 of Ptolemy's troops, and had increased his fleet by perhaps a hundred warships. This was the reason the battle was so important, quite apart from the conquest of Cyprus by Antigonos' forces; Ptolemy's armed and naval strength were both substantially reduced, and the naval balance was shifted decisively towards Antigonos. Ptolemy's fleet was very largely transferred into Antigonos' hands. Antigonos and Demetrios now had no naval challenger in the Mediterranean area.

Ptolemy is said by Plutarch to have got away with only 8 ships, but it seems likely that more survived. Many of the supply and transport ships were not taken, and even if he lost 120 warships (according to Diodoros), plus the 60 which had been in Salamis, he must have got away with 20 or 30 warships.[18] He will also have left some in Alexandria, but his sea power was, for the present, effectively destroyed.

Antigonos used the renown of the victory to proclaim himself king, and appointed Demetrios king as well. He thus set a precedent – that it required a victory to make a king, and if the king (or emperor) inherited his throne he had to justify himself quickly by a victory of some sort. Antigonos and Demetrios spent some time in celebration and coronation, and in establishing control of all parts of Cyprus. In military terms this was all an unfortunate delay, even if it was probably necessary in political terms. The Battle of Salamis is not closely datable, but must have taken place in the summer. (Demetrios left Athens in the spring – probably in March or April – and spent time in Karia, Kilikia, and northern Cyprus, and in attacking Salamis by land, before the sea battle.) All this celebration gave Ptolemy several months (late summer and early autumn) to recover from the defeat, recruit troops and ships, and prepare. It will have been clear to him that he would be the next Antigonid target.

The new expedition, in which both kings took part, was, as Ptolemy expected, against Egypt. This was the obvious move, but not very imaginative. Ptolemy was very seriously weakened by his defeat, and was not capable of mounting a revival for some time. It might have been more productive to return to Europe and try to eliminate Kassander, or even to attack Seleukos' lands – he was away in India and Baktria at this time. On the other hand, the conquest of Egypt would free Antigonos from the constant worry of an enemy on his southern flank. No doubt this, as well as personal enmity, was the justification. The expedition was huge: 80,000 infantry, 8,000 cavalry, 83 elephants, 150 warships, a hundred transports for the artillery.[19] The whole was assembled at Gaza by mid-October, and must have set out from North Syria in September. Ptolemy, if he had needed to be, was well warned.

The army marched by land from Gaza, taking the age-old invasion route along the north coast of Sinai, paced by the fleet offshore, which survived a minor storm on the way. Having reached the easternmost distributary of the Nile, the Pelusiac branch, the plan was for the fleet to land troops beyond the estuary, so opening the way for an easy crossing by the main force. Antigonos had trouble with his troops, who were no doubt not at all pleased by the long desert march. Some took an early opportunity to desert to Ptolemy – they may have been some of the men who had been captured in Cyprus. Ptolemy actively proclaimed the attractions of settling in Egypt, and Antigonos had to inflict nasty punishments to stop the rot. But Antigonos always had trouble holding the loyalty of his subordinates; even his nephews tended to desert him, and any Antigonid army which was beaten rapidly enlisted with the enemy. (Seleukos had collected 20,000 men in this way in a month or so in Babylonia in 312/311.)

The fleet made several attempts to put a force ashore. In each case it was found that Ptolemy had made quite adequate defence preparations, helped by the long delay in mounting the attack, by the memory of Perdikkas' attempt fifteen years before, and by the fact that there were not really many places where landings could be made. It had to be fairly near the Pelusiac estuary to be of use to the army, and away from swamps and other impediments. It cannot have been difficult to arrange the defence of all the likely landing places.

One landing was beaten off by missiles fired from the shore; another failed because the fleet became scattered and Ptolemy arrived with defensive forces at the landing place while the ships were still being collected. This landing was finally prevented by a storm in which three quadriremes and several transports were wrecked – but more importantly it was the high sea and surf following the storm which prevented any landing. Perhaps with the recollection of the failure of Perdikkas in the same attempt fifteen years before, Antigonos made no attempt to cross the river directly.

The other intention had been, having crossed the river, to dispute control of the Pelusiac branch by gaining control of the river itself. But Ptolemy's preparations had included the arming of small mobile river craft, which were more manoeuvrable in the current and the narrows and shallows of the river than the bigger sea-going ships – triremes and quinqueremes – which Demetrios had to use. The possibility of a march upriver to cross at Memphis, which had been the ultimate aim, was thus eliminated.

It is clear that Ptolemy had made good use of his time to make well-considered preparations to contest this invasion, but Antigonos had tens of thousands of men and over a hundred ships. He could certainly have absorbed many more casualties than he actually suffered. But several factors prevented him being too adventurous. He clearly could not rely wholly on his army after the desertions, and heavy casualties would obviously disenchant the troops even more – just as they had Perdikkas' soldiers. The navy's failure to make a landing, or to gain control of the river, was not decisive, for if Antigonos could have risked the crossing that would not have mattered. The sheer size of the expeditionary force was also a problem: camped east of the Delta in the desert it had to be supplied either by sea or by land from Palestine, and there were limits to what could be supplied over such a distance. A smaller force would have been easier to handle, more reliable, easier to supply, and much more likely to succeed. There are signs here of the same megalomania which had shown up in Alexander in his last years.

It is, of course, very easy to list the reasons for the failure of an enterprise afterwards, where small changes, minor successes, could well have produced a very different result. If the ships at the second landing attempt had not been scattered, a landing could well have been made before Ptolemy arrived to seize the landing place, and Demetrios might have got his troops ashore. But there was still a lot more to do even then, and by landing within the Delta Demetrios' forces would have controlled no more than a marshy island, with lots of other river channels still to cross. The problems of controlling the river, supplying the troops, and breaking down Ptolemy's resistance would thus all take time and casualties. And if Antigonos seemed to be winning in Egypt, Ptolemy's allies in Europe and Asia would take heed and mount distracting attacks.

It became clear that the expedition was in trouble. Antigonos consulted a meeting of his senior officers, who took counsel of their fears and recommended withdrawal, saving face by agreeing that it would be possible, and better, to make a new attempt next year, preferably when the Nile was low.[20] It was a defeat as significant as that which Ptolemy had suffered earlier in Cyprus, if less sanguinary, and Ptolemy took good care to publicize his defensive success, writing to his allies with exaggerated accounts – he took the title of king soon after to emphasise his victory. But Antigonos and Demetrios had not suffered any serious reduction in their power. Their casualties had been relatively few, given the huge size of their forces. It may be that in proclaiming his victory, Ptolemy led his allies to assume that Antigonos was no longer quite so dangerous. If so, they soon found they were wrong.

There was, however, the matter of Rhodes to be dealt with, so the prospective second assault on Egypt did not take place. Rhodes's refusal to assist in the Cypriot war clearly rankled with the kings, though it was scarcely an important element in the expedition. Antigonos had counted the island-city as an ally ever since he had commissioned the construction of ships in the shipyards in 315/314, and the city had loaned him ships in earlier wars. A refusal to assist in 306 was therefore a deliberate breaking of the alliance.

This is a crucial moment, therefore, in Antigonos' conception of his power. He had been reasonably scrupulous in treating Greek cities with consideration ever since his proclamation in 314 that his policy was to secure their freedom (as against Kassandros, in particular). But the success of this policy depended on those cities becoming, and remaining, his allies. Since he had the predominant power and prestige, alliance with Antigonos meant that the ally was expected to contribute to his expeditions in the way Demetrios had clearly expected Rhodes to do on his way to Cyprus. The policy had worked well for ten years – Athens, for example, was now an ally, and had contributed ships and crews to the Cypriot expedition. But Antigonos clearly argued that once one ally got away with a refusal, the rest might well follow its lead. Rhodes's refusal to assist Demetrios was a body blow to Antigonos' whole political system, though the city may well not have realized it at the time.

Rhodes' defiance, and its apparent preference for Ptolemy because Egypt was the city's main trading partner, put it in the ranks of Antigonos' enemies. And if so it was clearly dangerous; it was also in a similar strategic position to Cyprus. Geographically it was a large island commanding the narrow sea lanes between the Asian mainland and Crete. Not only that but it controlled part of the nearby mainland, the Rhodian *peraia*, and the islands of Karpathos and Kasos between itself and Crete. In recent decades the city had built up an excellent fleet, small but professional, which had already seen off an attack by Perdikkas' brother Attalos with a large fraction of the fleet of Alexander. To Antigonos, therefore, Rhodes lay across his main imperial line of communication and its new hostility made it very dangerous.

The events leading to the outbreak of the fighting are a case study in escalation. Antigonos sent a second embassy, while the fighting was going on in Cyprus, asking that Rhodes resume its alliance and send ships after all. The city refused, and Antigonos identified its trade with Egypt as a useful pressure point. He sent a commander with some ships to intercept Rhodian ships trading with Egypt, and confiscate their cargoes. This was, given his war with Ptolemy, a reasonable precaution, but when the Rhodians drove his ships off, he took it as a hostile act, which it was. This was only a minor incident, for neither party was using very much force. Rhodes was only a very minor naval power at the time, disposing of relatively few ships, and Antigonos was more anxious to bring Rhodes into the alliance once more than to punish the city. But it was clearly recalcitrant, and Antigonos declared that the Rhodians had started the fighting, though that was clearly disputable. Exactly when this sea scramble happened is not clear, but late in 306 is a likely time, while Antigonos was making his grand assault on Egypt.

Rhodes, finding itself likely to be attacked, attempted to pacify Antigonos by voting him honours, but significantly did not accept that the alliance should be renewed. Instead the Rhodians declared that the city had treaties with Ptolemy, which should not be broken. Antigonos was rightly angered by this, no doubt remarking with some force that this is exactly what Rhodes had already done, but with him.

Demetrios was despatched with a substantial force to bring the city to reason. He was greeted, apparently before he arrived in the area, by another Rhodian embassy, agreeing to resume the alliance, and to fight Ptolemy after all. In face of this wriggling and the Rhodians' manifest unscrupulousness, Demetrios demanded that a hundred hostages be handed over, selected from the rich, and that his fleet should be accepted in the harbours of the island. This would, of course, deliver the city into Antigonos' power, but the city's diplomatic history, and its clear lack of consistency, must have convinced Demetrios that something more than a paper agreement, even with oaths, was required to pin the city down.

Rhodes refused these terms, apparently not even attempting to negotiate them down to something more acceptable. It is clear that the city's policy had been both unprincipled and stupid, and that Demetrios had been driven to war by the city's evasiveness. After the victory in Cyprus he must have felt that Rhodes was a much easier problem; and after the withdrawal from Egypt – the defeat, Ptolemy was calling it – he may well have felt that another victory was needed to restore the prestige of the two kings. But by attacking a city intent on keeping its liberty, Antigonos and Demetrios were revealing that their own 'freeing the cities' policy was one which was really aimed at increasing their own power at the expense of cities which were already free.[21]

As such Rhodes and Demetrios came to blows. Demetrios had overwhelming power at sea, supposedly arriving with 200 warships of various types, and 170 transports. On the ships he is said to have brought an army of 40,000 men.[22] These figures must provoke some scepticism, both in themselves, and in regard to Demetrios' purposes. The 200 warships is just about possible, given that he had almost as many at Salamis and had captured over a hundred more from Ptolemy and Menelaos, and using 200 in the west would still leave a serviceable number to guard the east, in case Ptolemy broke out. But this total of ships was quite unnecessary for a war against Rhodes, as was an army of 40,000 men. It very much looks as though Demetrios was really intending to go elsewhere, and that Rhodes was something he aimed to deal with on his way. The obvious target would be Athens and a new war against Kassandros. He cannot have expected Rhodes to resist for very long.

But the Rhodians did resist, enduring a siege lasting a year, which taxed everybody's ingenuity and originality and endurance to the limits. The Rhodians, apart from being attacked by this great force, also found a great crowd of pirates and plunderers hanging around Demetrios' forces, waiting to pick the city clean. These ships actually proved to be vulnerable to Rhodian raiders, which were ships of the *hemiolia* type, smaller and more lightly manned than triremes, but swifter. These raiders did little real damage, and Demetrios was not inconvenienced – he probably did not care about the plunderers – but the Rhodian ships were clearly able to get out of the harbour, and also were able to bring in food supplies, a factor which much extended the siege. The Rhodian naval vessels, some quinqueremes and more quadriremes, are conspicuous by their absence in this war; presumably, given Demetrios' large fleet, the Rhodians kept their ships in harbour, since they could not possibly succeed in a battle. They did use three ships, described as their strongest, which are assumed to be quinqueremes, against some ships carrying Demetrios' siege engines.[23]

Despite this, the siege was not really an exercise in seapower. Once Demetrios' army was on shore it was a land fight. The sources for the siege are fascinated by the great siege machines Demetrios used, but are also generally sympathetic to the Rhodians. The result is that only a few episodes are known at all well. In the end Demetrios was receiving reports of a serious threat to Athens from Kassandros' forces, and it was known that Seleukos was returning to the west from his great eastern expedition. Ptolemy had been active in sending food and military supplies to the Rhodians, and both Lysimachos and Kassandros had also sent in food. Antigonos' enemies were gathering once more, and the siege was becoming an embarrassment to everyone. The Athenians tried to mediate, so did the Aitolians – both of these were allies of Antigonos. Ptolemy promised more aid, but also urged the Rhodians to seek peace, and held back delivery of the aid for the moment.

In the spring of 304 Ptolemy, Kassandros, and Lysimachos renewed their alliance, and at last Antigonos intervened personally, instructing Demetrios to settle. On the other side the Rhodians must have realized that the help they were getting from Kassandros, Lysimachos, and Ptolemy was only keeping them going, and that it was in the interests of these kings to see Demetrios pinned down at the siege as long as possible – that they were being used by their allies, in other words. At this opportune moment an embassy from the Aitolian League offered to mediate, and so terms were agreed, in the end rather easily. The Rhodians gave up a hundred hostages and agreed to be Antigonos' allies in all conflicts except against Ptolemy.[24] Demetrios, after a year's siege, had gained his point, but at the cost of considerable expenditure of resources, and was widely seen to have been in effect defeated.

The attacks on Egypt and Rhodes had not enhanced the power or the prestige of Antigonos. Both episodes were widely seen as defeats for him, though Rhodes remained quiet for several years, while its hostages were in Antigonos' hands, and Ptolemy's power was so reduced by the defeat at Salamis that he was also quiet. The great victory at Salamis was still having its effect. The earlier victory at Athens was threatening to become undone, and this gave Demetrios the opportunity to score another success.

Kassandros had been nibbling away at Antigonos' allies in central Greece ever since Demetrios had sailed for Cyprus. He had taken control of Chalkis and Eretria in Euboia, had captured at least two of the Athenian frontier forts and the island of Salamis, and by 304 he was making direct attacks on the city itself. At one point the walls were breached, but the city's cavalry blocked the incursion. Yet it was clear that the city would fall in a relatively short time unless it was relieved.

This situation was one of the factors which helped Antigonos decide to cut his losses at Rhodes. Demetrios gathered up his ships and men and made another of his voyages. He now had 330 ships, so either he had lost forty at Rhodes, or he had sent some of them elsewhere (or he did not have 370 in the first place). Clearly fully aware of the position in Greece, he ignored Athens and chose to land at an unexpected place, unguarded by Kassandros' forces, at Aulis in Boiotia. This instantly isolated Euboia and the cities Kassandros held there fell at once. Demetrios was also in a position to strike into Boiotia, and the Boiotian League at once capitulated. Along with his alliance with the Aitolian League, this put Demetrios in effective control of all central Greece from Megara to the borders of Thessaly, and cut off Kassandros' forces in Attika from his base in Macedon. It had been a most effective use of his sea power.

Kassandros' forces were driven back, and Demetrios seized the city of Herakleia in Malis, which guarded the northern end of the Thermopylai pass. Turning south he then retook the Athenian forts, and presumably Salamis Island as well.[25] The Athenians awarded him still more honours, though much of the work of resistance to Kassandros' attacks had been their own. Demetrios moved on past Megara (which he had taken in 307) and captured Kenchreai, the eastern port of Corinth. Next year from that base, and using some ships he had acquired or sent into the Gulf of Corinth, he took Sikyon from Ptolemy's troops, and then Corinth from Kassandros', both in swift and night-time attacks. Further conquests in the Argive area followed. By 302, when he and his father organized their Greek allies into a new League of Corinth on the pattern of that of Philip II and Alexander, most of Greece between Thessaly and Sparta was in that alliance.

These expeditions commanded by Demetrios earned him the cognomen of *Poliorketes*, 'city-taker', though he had been as much a failure as a success in his sieges. From a strategic and tactical point of view, however, it is the use made of the sea power he and his father controlled which is the most revealing. They began with a fleet which was by no means overwhelming in size. In Cyprus Demetrios had fewer ships than Ptolemy and he won at Salamis by his dispositions, placing his fleet between the two smaller sections of Ptolemy's. At a strategic level, he had now done the same, between Kassandros' fleet and that of Ptolemy. Once Ptolemy was beaten he could turn back to Greece or go against Egypt as he chose. The two kings (as they now were) chose the second. It made some strategic sense, but failed in the execution. Then Demetrios turned back to Greece, where his landing at Aulis at once gave him the strategic initiative, which he used successfully to conquer much of the country in the next two years.

In all this his sea-strength was decisive. Its most obvious use was to escort armies to points of attack – to Athens, to Cyprus, towards Egypt, to Rhodes, to central Greece. This was also what Ptolemy was using his fleet for before his defeat at Salamis, and the consequences of an army afloat without an escort became obvious when many of his troops were captured after the battle. The transports were obviously vulnerable and it is clear that moving armies by sea was a complicated business.

This was the broad strategic use made of the navy. In tactical matters Demetrios showed a considerable imagination. The surprise of the Peiraios in 307 was the decisive action in bringing about the fall of Athens; a conventional landing elsewhere in Attika would have given the Macedonians in command in the city plenty of time to organize their defences, cow the hostile population, and to get word out to summon reinforcements. Instead the port came quickly under Demetrios' control, and by proclaiming freedom he brought many of the Athenians to his side. The same tactic was used on a much greater scale three years later in his landing at Aulis, which brought the collapse of Kassandros' position throughout central Greece. These were tactical master strokes which had great strategic results.

Demetrios was the only commander to win a major battle at sea between the Athenian defeats at Abydos and Amorgos in 322 and the battle of Kos (probably in 260). One reason, of course, is that after Salamis no other power had the ships to challenge him. But it is noticeable that even when Ptolemy and Antigonos were roughly evenly matched, in the years before the encounter at Salamis, no battles had resulted. Ships were expensive, and to risk them in a battle was clearly not worthwhile without the assurance of victory – but outnumbered fleets would not normally allow themselves to get into a position where they could be attacked. Salamis was fought because something more valuable than the fleet, the island of Cyprus, was at stake. In other cases this was not so, and battle was avoided.

In other words, the value of these fleets was more for their tactical and transport uses than as a means of fighting enemy fleets. The 'fleet-in-being' concept, to use a later term, was uppermost in the minds of commanders and rulers. There was no attempt to establish total control over the sea, so that no ships could sail except by leave of the controlling power, though control of coasts had that effect, and certainly so in the case of Syria.

The use of small squadrons was much preferred. Kassandros at Oreus, the Rhodians during the siege, Polykleitos on the Kilikian coast, were all locally effective actions using relatively small squadrons of ships. Similarly Antigonos attempted a sort of blockade to stop Rhodian ships trading to Egypt, but this was as much a move in the diplomatic game as it was a blockade, and in its latter role it is unknown how effective it was, though it certainly angered the Rhodians.

This difference between the use of a whole fleet and of smaller squadrons was a result of the type of ships employed. Galleys are highly manoeuvrable, and a well-manned galley with a fresh crew could move fast, at least over a short distance. Unless, as at Salamis, the whole harbour entrance could be blocked, it was always possible for a galley to escape (and even at Salamis, the sheer number of Menoetios' ships forced their way out). So, except in unusual circumstances, a close blockade was not possible. Even Demetrios' boom and the capture of the mole at Rhodes did not stop Rhodian ships getting out.

So it was in small-scale actions that sea power had its main effect. The landing at Peiraios and the use of ships to surprise Sikyon are examples of the clever use of small squadrons, but this could only succeed if these small landing squadrons were backed by, and protected by, the fleet as a whole. At Oreus, the removal of Medeios' main fleet had made it possible to counter-attack the small Antigonid force. At the Peiraios the surprise landing had to be quickly reinforced, and at Sikyon the seaborne landing was part of a more elaborate attack which included assaults by land. And these moves did not always work, as the failures of the landing attempts in Egypt demonstrated. Resolute and swift opposition at a landing point could defeat such attacks. Ptolemy tended to use his fleet in the same way, as in his cruise in 309/8, when he seized several bases in Asia Minor and Greece – but without support the Greek places were easily lost as soon as Demetrios attacked them.

These campaigns by Demetrios therefore demonstrated that sea power was exercised above all in relation to the land. There was no 'blue-water' strategy, no command of the ocean, no long-term blockade of the enemy ports, and large-scale battles were extremely rare. But fleets were ideal vehicles for surprise attacks, for landings, for outflanking enemy forces. Command of the sea in the decades after Alexander meant the ability to move large armies at will by sea, and so also the ability to deny that capability to enemy forces; second, it meant being able to put forces ashore, and support them, in order to assist or begin a conquest. A subsidiary purpose, which Ptolemy pursued, was to create a series of bases by which Egypt could be kept in contact with Greece and Asia Minor.

Chapter 5

Demetrios the Sea-King and the Super-Galleys

In 302/301 Antigonos' enemies finally cornered him. Seleukos brought a great cavalry and elephant force from the east, Lysimachos brought a great infantry force from Thrace, Kassandros blocked Demetrios' attempt to invade Macedon and sent half of his army as reinforcements to Lysimachos. Demetrios in turn brought his army to Asia to join his father in Asia Minor. The two great armies manoeuvred over the winter of 302/301 and in 301 finally met in the Battle of Ipsos. Demetrios and the Antigonid cavalry were enticed away and blocked from returning by Seleukos' elephants; Lysimachos' and Kassandros' infantry ground down Antigonos' own infantry in a great killing match, whose fatal casualties included old Antigonos himself.

In this campaign sea power played its part. Some of Kassandros' troops were drowned when their ships were caught in a Black Sea storm. They were crossing in the north in part because Antigonos' ships controlled the easier crossing at the Hellespont. Demetrios' fleet controlled the Aegean, in the sense that he was able to move his forces without any hindrance, to Thessaly, to Ephesos, to the Hellespont and the Bosporos, starting from Athens, evidently his main base.[1]

Lysimachos was able to cross the Hellespont with his large army at the start of the campaign, but when Demetrios sent thirty ships to the Bosporos this was enough to gain control of the crossing, so that Kassandros' force, which was sent to assist Lysimachos in Asia, had to cross from Odessos to Herakleia, a voyage in which some of them were caught in a storm and wrecked.[2] At the end of the campaign, after defeat in the great battle, Demetrios fled from the battlefield back to his ships.[3]

The political result of Antigonos' defeat was a carve-up of his kingdom, with each ally taking his share; Demetrios was left with the fleet and only those places he could control from the sea: some cities on the Ionian coast, Corinth, islands in the Aegean, Cyprus, and Tyre and Sidon in Syria. Athens had become weary of him by now, of his wild behaviour and his arrogance, and shut him out, though he was allowed to take his ships away with him.[4]

The usefulness of his fleet was demonstrated in the next years. Demetrios had sufficient ships to command the sea, in the sense that he could land where he wished. First he took his ships and his army (he had escaped from Ipsos with 9,000 men) to the Chersonese to ravage Lysimachos' lands, and so gained some revenge and also replenished his treasury. He then sailed to Syria, where Seleukos, who had acquired only North Syria as his share of the spoils, was thoroughly dissatisfied and was just as politically isolated as Demetrios himself. The two formed an alliance, with Seleukos marrying Demetrios' daughter Stratonike. While he was in the east Demetrios seized control of Kilikia, which had been allocated to a brother of Kassandros; he also acquired a substantial treasure which had been held in the fortress of Kyinda. A further bonus was that Seleukos reconciled Demetrios with Ptolemy, who handed over a daughter to be Demetrios' next wife.[5]

The friendship of three such men could not last. Seleukos demanded Kilikia from Demetrios, perhaps as Stratonike's dowry; he was refused, and asked for Tyre and Sidon instead, cities also desired by Ptolemy. Demetrios again refused, and reinforced his point by a raid into Ptolemy's territory; the marriage to Ptolemais did not take place for thirteen years. Demetrios had by this time gained as much as he could from these exchanges, and returned to the west, where Athens was once more in crisis.

Before he could do much his fleet in the Aegean was struck by a storm; most of the ships he had with him were wrecked and many of his men drowned. He collected such men as he had and invaded Attika, capturing part of the land and imposing a food blockade on the city. (His method was to capture one grain ship, then hang the captain and the supercargo; this understandably scared off the other grain ships.) He sent out men to collect such vessels as he still controlled, which had evidently been dispersed to other ports, no doubt as an ever-present reminder of his authority. Ptolemy, evidently hearing of, and being encouraged by, his misfortune, arrived to try to assist Athens. But Demetrios' messengers had brought together his ships into a new fleet by that time. Ptolemy's fleet of 150 ships was outnumbered by double that, and he withdrew.[6]

How many ships Demetrios lost in the storm is not known, but it was clearly a considerable fraction of his total fleet strength. The 300 ships his men brought together afterwards came from places he controlled in the Peloponnese and from Cyprus, and presumably from other places he held as well – in Tyre, Sidon, the islands, and Ephesos. In 304 he had commanded 330 ships; in 305 at Rhodes he had 370; after Salamis in 306 he had 290, and there were probably more which Antigonos had kept in Syrian and Aegean ports. There had presumably been more new building in those years as well, so 300 is a reasonable number for him to control even after the storm. He obviously owed the survival of a fleet of overwhelming size to the fact that it had been dispersed, though this may not be taken as the reason for its original dispersal. By collecting them from all around his separated lands, however, he was now leaving some of these distant places less well protected and so vulnerable.

Ptolemy had been busy rebuilding his strength in the past ten years. He had been unable to join his allies in the Ipsos campaign, but he had helped himself to part of the spoils nonetheless. He secured Palestine and part of central Syria as far north as the Eleutheros River; Demetrios held on to Tyre and Sidon, and Seleukos took the land north of the Eleutheros. (Seleukos actually claimed the rest of Syria, but did not have the strength to challenge Ptolemy for it.) Ptolemy may well have acquired some of Antigonos' ships in the harbours of Palestine, but it is obvious that he had also been building new ships for himself as well. Assuming he had escaped from the aftermath of Salamis with thirty or forty ships, and had previously left some in Egypt, he must have built at least a hundred more since then. (And he had done so without access to the forests of Cyprus or Syria; so much for his need for Cypriot forest resources.) Given the treelessness of Egypt, the timber had either come from the stocks he had built up, or was imported. Rhodes is the obvious medium for this trade. His continued control of ports in southern Anatolia, such as Phaselis and Korakesion, both close to forested hills, meant he also had access to such supplies directly.

Demetrios frightened Ptolemy's fleet away, besieged and conquered Athens, which he garrisoned – the old 'freedom' policy had disappeared, except in name – and then he went on to campaign into the Peloponnese. But all this took time. Athens had held out for months, until starvation forced surrender. Just as during the siege of Rhodes, this long siege allowed his enemies to consult and prepare and arm. They were alarmed at his victories, and enticed by his new weakness, and scared by his ambition, and all of them attacked him together. Ptolemy and Seleukos seized Cyprus and Kilikia respectively. Seleukos had merely to march his forces across the mountains to take Kilikia, but Ptolemy had to use his fleet. Demetrios had apparently withdrawn his Cypriot squadron to the Aegean, so Ptolemy won most of the island with ease, though he then had to besiege Salamis for a time. In Ionia, Lysimachos seized the cities Demetrios had held, including Ephesos. This was one of the results of his concentration on Athens and his naval concentration in the Aegean area. In gaining Athens he had lost most of those places which made him a sea-king.[7]

He rebounded almost at once. Kassandros had died before the siege of Athens, and his eldest son soon followed; his younger sons then quarrelled. In stepped Demetrios, eliminated the two boys, and took over Macedon and its kingship. At once he displayed those qualities which had first attracted the Athenians to him and then sickened them, and had repeatedly alarmed his fellow kings. Arrogantly he neglected the Macedonians, eagerly he built up his military and naval strength. He is said to have built up an army of 98,000 foot and 12,000 horse, and a fleet of 500 ships. These figures are probably exaggerated, but, in terms of building the ships, he did have control of the shipyards at the Peiraios,

Corinth, Chalkis, and in Macedon; and he began with the 300 ships he had collected earlier. A total fleet of 500 is quite possible, but like his father's boast of building 500 ships in 315, it was probably an aspiration rather than an achievement.[8] The sheer number of men needed for such a fleet makes it highly unlikely that it was ever in existence – even if the ships were only triremes, the rowers, officers, crews, and marines would number over 100,000 men. It does not seem likely that Demetrios, out of the resources of Macedon and his Greek cities, could have afforded to pay and equip well over 200,000 soldiers and sailors.

The Macedonians, annoyed at his casual attitude to ruling and kingship, in effect seceded from under him, going over to Lysimachos and Pyrrhos of Epeiros, who, encouraged by the other kings, invaded from east and west respectively. Ptolemy arrived in the Aegean with his fleet, and persuaded some of Demetrios' cities to change sides, and he was shut out of Athens again. Enraged, he sailed to Asia Minor to attack Lysimachos, so effectively abandoning the sea. He left his son, Antigonos II Gonatas, in Greece with most of the ships, and the ships which had carried him to Asia were left at Ephesos.[9] He marched through Anatolia, but was never able to bring about the battle he wanted. Eventually he was pushed over the Taurus Mountains and was captured by Seleukos, who imprisoned him and let him drink himself to death.

Demetrios had maintained himself as a great power for sixteen years, from the Battle of Ipsos in 301 to his capture in 285, largely on the basis of commanding a fleet larger than anyone else's. No doubt all his many enemies were profoundly relieved when he was captured, though, since Seleukos deliberately held him prisoner, with the implied threat of loosing him once more, it was only when he died that real relief was granted them. His son Antigonos still commanded part of the fleet and inherited his aspiration to rule in Macedon, but he was burdened by the memory of his father's conduct and extravagance, and it took him ten years of effort before he was firmly seated on the Macedonian throne. So the capture and death of Demetrios was a distinct conclusion to a phase of affairs, all the more so in that Ptolemy, Lysimachos, and Seleukos all died in the next two years. By 281 a new generation was in power, with different aims.

Behind all the heavy political events between 301 and 281, however, naval developments had been just as extraordinary. Naval power depends in part on the number of ships deployed and the size of the fleets assembled; it also depends on the quality of the vessels in those fleets. The size of Antigonos' and Demetrios' fleet between 315 and 302 has been noted already in this chapter. Demetrios' building programme in 290–288 was aimed, we are told, at producing a fleet of 500 ships. During the events of 288–285 that fleet, which was probably nowhere near 500 strong, broke up. Some ships will have been left in Macedon, to be taken over largely by Lysimachos; some will have remained in the harbours of Greek cities, principally Athens, which rebelled again. Some were taken by Demetrios to Miletos, and then some of these moved away to Kaunos. We do not know, however, how many ships there were in each of these sets, but one major portion was the Kaunos group, and this became the inheritance of Antigonos Gonatas.

Meanwhile Ptolemy developed, as noted, his new fleet, of which 150 ships were in the force which attempted to assist Athens in 297/296. In the crisis of 288, when the new coalition formed against Demetrios, Ptolemy sent a 'great fleet' into the Aegean once more, a term which rather suggests a bigger force than the 150 he used earlier. Since this fleet was intended to challenge that of Demetrios, it is evident that Ptolemy had continued to build. Having taken Cyprus in 294 he had increased resources for doing so.

When Demetrios fell from power, far off in the Syrian mountains, he left his son in charge of his assets in Greece. He had left his main fleet at Miletos, but, once he had gone off into the interior, Lysimachos' commanders recovered those cities he had taken, including Miletos. At that point the fleet there split up. There seems to have been no question of Lysimachos gaining control of the ships, and it seems that the break-up was peaceful. Part of the fleet went off to Kaunos and remained loyal to Demetrios, and eventually, on Demetrios' instructions when he was captured, went over to Antigonos, who was in Athens or thereabouts. The other part of the fleet was partly from Phoenicia and was commanded by Philokles, king of the Sidonians. He took himself and his ships into Ptolemy's service, and became Ptolemy's naval viceroy in the Aegean.[10] The negotiations over all this must have been complicated, but the Phoenician cities, Tyre and Sidon, also surrendered to Ptolemy at the same time, and this was possibly on Philokles' orders, or at least at his recommendation. Philokles soon sailed to Kaunos and captured the place for Ptolemy, though it seems that Antigonos' ships (formerly Demetrios') were no longer there.[11]

Philokles' actions had wider effects. With Tyre and Sidon now in his control, Ptolemy had rounded out his province of Koile Syria, and had dashed the last hope of Seleukos of acquiring any of Syria south of the Eleutheros River. Demetrios must have been chagrined also, since the desertion of Philokles (as he would have put it) and the loss of Tyre and Sidon and Kaunos, happened while he was still on campaign. But from the point of view of sea power, Philokles' switch of loyalty proved to be the decisive transfer of the command of the eastern Mediterranean seas into Ptolemy's hands. It is not known how many of Demetrios' ships followed Philokles and how many went to Antigonos, but by adding a substantial fraction of Demetrios' ships to Ptolemy's fleet, and adding in those ships which Ptolemy will have built in the last few years, it is clear that he had now acquired the preponderant naval power.

There remained, however, a number of lesser fleets in the Aegean area, which was at the extreme end of Ptolemy's reach. Antigonos had one of these fleets, and he held several important port cities – Corinth, Peiraios, Chalkis – which were well fortified, productive of taxation, and had important shipyards for building warships. He had inherited only part of Demetrios' fleet, but that could be a hundred ships or more. Lysimachos had effective control of a group of relatively small fleets belonging to cities along the Propontis and the Pontos – Herakleia, Kios, Chalkedon, Byzantion – which formed themselves into a Northern League after his death and could at that time put a fleet of considerable force to sea. Herakleia probably inherited a substantial part of Lysimachos' fleet when he died in 281. Lysimachos had no doubt also collected some of Demetrios' ships in the harbours of Macedon and Thessaly, particularly at Demetrios' new city of Demetrias in Thessaly, but by 285 Ptolemy had the edge on everyone else, thanks to Philokles.

The types of ship employed in these fleets, as described in the accounts of the Battle of Salamis, were mainly triremes, quadriremes, and quinqueremes. Despite later developments of bigger ships, these remained the main categories throughout the Hellenistic period, and beyond. Each had its own qualities, but it never seems to have been possible to combine all the good qualities into one vessel. The trireme was, in this company, small, but it was swift and manoeuvrable. The quadrireme was heavier but almost as manoeuvrable. The quinquereme was by no means as easy to move and manoeuvre as the others, being considerably heavier, but that made it all the more formidable in battle; it was quinqueremes which were the preferred form in most fleets from now on. These became the normal working warships, the ones which made up most of the ships in the fleets, which went on patrol, and which did most of the naval work.

The competition for naval power had another aspect, besides the size of the fleets. New types of ships, bigger than quinqueremes, were built, designed to enhance the effectiveness of those fleets. At Salamis, Demetrios had shown very clearly just how useful such new types were in the line of battle; so anyone with a fleet needed some of them. Their advantage, of course, lay, like the quinqueremes, in their size and weight and height, but they were expensive to build and employed very large crews. These ships were generally identified by their oarage in numbers, like the term trireme and so on. The sixes and higher were consequently few in number. But they were also different, for they were built for other purposes, not always necessarily naval.

Later historians claimed that some of these bigger ships were built, or at least ordered to be built, by Alexander, but this seems unlikely. Antigonos in his new building programme in Phoenicia and Kilikia in 315/4 is said to have built nines and tens, but the fact is that these ships are never heard of again, and it seems very likely that, like Alexander, he was being credited with them retrospectively, simply because he ordered a lot of ships to be built. Sixes, however, are known to have been built in Sicily by Dionysios I, though this is not particularly well attested.[12] And Demetrios had ten sixes on his victorious left at Salamis, and seven sevens as well.

The names given to the ships describe the number of rowers. It was normal to propel a galley with two or three banks of oars, rowed by oarsmen in files, and it is the number of files which gives each class its name. The trireme, always the basis from which the discussion must proceed, had three oars in each set, each pulled by one man, so there were six files of men the length of the ship. Quadriremes had two oars manned by two men each; quinqueremes had three oars, two of them pulled by two men, and the third by one. The size of each class of ship had to be greater, of course, as the number of files increased.

Going up from a trireme to a six was relatively straightforward, for a six can be considered to be just a trireme with two men rather one man to each oar, though obviously it was bigger. The numbers of the rowers were growing, however, from the trireme's 170 to the quadrireme's 176 to the quinquereme's 300 – the quinquereme was thus a major change. Going beyond this size was more difficult, for the ships became heavier and required more men with each manipulation of size and shape. A seven, of which Demetrios had seven at Salamis, was a bigger ship than a six, and had about 400 oarsmen; it had three banks of rowers, one with three men pulling and the others with two. In some navies the six was the biggest ship built, and it was used as a command vessel, just as the quinqueremes had been in the navies of the Phoenician and Cypriot kings.

These larger ships, up to tens, were certainly used in battle, but anything above a ten was never seen in action, though they were certainly built. Nines and tens seem to have been bigger versions of the quadrireme, with two banks of oars pulled by five and four men in a nine, and two sets of five in a ten. These were clearly huge ships, and in particular they were also higher than others, giving the fighting men on board an advantage over those in the lower ships. But the higher the gunwale of the ship the wider the beam necessary to lower its centre of gravity and allow it to float. The ships of this size were too heavy to be easily manoeuvred, and they obviously took some time to build up speed; they were effectively floating fortresses around which the battle was fought, for the smaller vessels could fairly easily avoid them. That is to say, the bigger ships, above sevens, at least, were built less for their practical battlefield use in the traditional manoeuvres of battle than for their sheer size. They carried a larger force of marines than the smaller vessels, who could be used to board enemy ships. There was also great prestige attached to having them.[13]

Demetrios' success at the Battle of Salamis with his sixes and sevens seems to have been the trigger for the naval arms race which followed: a competition between the kings to build more and bigger ships. Or perhaps it was just as much a matter of Demetrios deliberately forcing the pace and others looking to catch up. By 301 he had a thirteen, which he left at Athens while campaigning in Asia Minor, and which is one of the ships the Athenians let him take away when they refused to let him re-enter the city. This ship was later taken to Rhosos in Kilikia where he entertained Seleukos on it and where the two kings made their alliance.[14] That is, it was not used in any fighting, but was a considerable diplomatic asset.

Lysimachos was the ruler who attempted to compete first. He controlled the city of Herakleia-on-Pontos and there he had an eight built. This ship, called the *Leontophoros*, was famous for its beauty. It has been suggested that this was merely a sixteen, built as a double-hulled catamaran, but such a ship would hardly be regarded as beautiful to eyes accustomed to admiring triremes; it is best seen as a very large, long and wide, vessel, and so retaining the grace of the smaller ships; it had a crew of 1200 men, sailors, oarsmen and soldiers. It is reckoned to have been about 110 metres long, almost the size of three triremes, and this made it unusual.[15]

Demetrios also built fifteens and sixteens, which were also admired, but rather for their size than their beauty. Lysimachos once asked Demetrios to show them to him, and stood on the shore as Demetrios sailed past in his monster ships while everybody marvelled.[16] This is a scene which surely encapsulates the very purpose of these vessels. When the *Leontophoros* and Demetrios' fifteen and sixteen were built is not clear – some time between 302 and 286 is the best we can do, for it was only in that time frame that the two kings were in direct naval competition. Probably Lysimachos' ship was first, being to a degree experimental; it is perhaps best to think of the bigger ships built in the later part of that period, when Demetrios was busy expanding his navy, having gained the Macedonian kingship. In which case, of course, he lost control of them very soon after they were built.

Demetrios himself, so it seems, was both the organizer and in many ways the architect of the ships. They would certainly not have been built but for his insistence. And he is described, in Plutarch's biography, as visiting the shipyards 'in person, showing what was to be done, and aiding in the plans'.[17] The actual plans were presumably made by a naval architect, but it was Demetrios whose ideas ruled. The same may well be said of Lysimachos' *Leontophoros*. Unlike Demetrios' monsters, however, *Leontophoros* was used in battle, successfully, in a fight between Antigonos Gonatas and Ptolemy Keraunos in 280.

By the time the generation of Alexander's contemporaries died off, therefore (between 283 and 281), there were several of these huge ships at sea, but the battle fleets still consisted mainly of quinqueremes and smaller ships, with a few ships up to sevens and even tens acting as command centres and, if possible, used in the fighting. The splitting of Demetrios' fleet saw his biggest ships, the thirteen, the fifteen, and the sixteen, go to his son Antigonos, probably because they were left in the Greek ports when he went off to Asia. Philokles' part of the fleet, going to Ptolemy, had the smaller vessels, though these probably included some of the sixes and sevens. The ships in that category at Salamis had been added to Demetrios' force when he stopped at Kilikia before invading Cyprus; and the sevens are explicitly stated to have been Phoenician.

Given the competitiveness of the kings, it is thus not surprising to find the successors of Demetrios and Ptolemy at daggers drawn. Antigonos II and Ptolemy II were as earnest and lifelong enemies as their fathers had been, and one element of their enmity lay in their navies. Given that Antigonos had the three great vessels, it was clearly necessary for Ptolemy II to catch up with his enemy and then surpass him in this area. This was not in fact very difficult to achieve, for Antigonos for a long time had relatively few resources with which to enter such a competition.

Antigonos had ruled only a few Greek cities when his father was captured. For eight years this remained his condition. His aim during that time was to succeed his father as king in Macedon, but first Lysimachos controlled that kingdom and then it was invaded and destroyed by bands of Celtic Galatians from the north. In the midst of all this Antigonos had to fight a battle at sea against a competitor, Ptolemy Keraunos, who had acquired the use of Lysimachos' old *Leontophoros* in his alliance with the Northern League. That ship virtually won the battle for Keraunos, who seized the Macedonian throne soon afterwards only to be killed by the Galatians. Antigonos had obviously not used his biggest ships in the fight. Eventually, in 277, Antigonos was accepted as king in Macedon, having cleared the Celts out by the sensible device of hiring one band to fight the rest. But Macedon in the meantime was ruined. So, for perhaps two decades after Demetrios' death, Antigonos' navy remained essentially what he had inherited, probably with very little new building being done.

In that time, Ptolemy II, who succeeded his father in 282, constructed a fleet which no-one else could match. His maritime imperial system, and the monster ships he made, will be discussed in the next chapter.[18] He went so far as to have a twenty and two thirties built, which outdid everyone else by a long way. He was eventually outshone by a forty built on the orders of his grandson, Ptolemy IV. It is well described and would seem to have been of the double-hulled catamaran type, planked over to provide a huge floating platform. It is likely that the earlier monsters were also of this type. None of these enormous vessels was really of any use except to feed the vanity of the kings and their naval architects. A sixteen was later described as almost unmanageable; twenties and thirties would have been useless except as prestige items; the forty never went to sea, being confined to agonizingly slow movements on the River Nile.

This arms race, or prestige race, was, of course, enormously expensive, but these kings were drawing on the wealth stored up by the defunct Persian power and released by Alexander's conquest, and they were also adept at squeezing taxes out of their subjects. But they were also spending that money, so that it was also being circulated in the best economic theory style. It is clearly one of the major elements in the expansion of the general economy which was evident in the Hellenistic world. In terms of sea power, it was Ptolemy I and II who emerged as the dominant rulers once Demetrios' fleet had disintegrated.

Chapter 6

The Ptolemaic Sea-Empire

The competition between the kings had by 283 produced three major navies in the eastern Mediterranean, but one of these, that of Ptolemy I, was substantially the most powerful. Ptolemy died next year, and his son and successor further built up his fleet's overwhelming power. Just how strong it was is seen in the account of the fleet of Ptolemy II preserved as an extract from some unknown source by Athenaios, compiler of a wide variety of extracts on many subjects much later. He seems to have culled his extracts with some care, and it is likely he had preserved reasonably accurately the composition of the fleet as his source recorded it.

After a slightly uncomfortable start to his sole reign, Ptolemy II, called Philadelphos, developed a policy in foreign affairs which relied in part on vigorously and grandiloquently displaying his power, and this was a major deterrent to his enemies. It did not work all the time, and in particular he found he had to fight the Seleukid kings more than once to keep control of Koile Syria, as well as fighting a long war in the Aegean over control of Athens. He had enormous resources, extracted through taxation from the long-suffering peasantry of Egypt, and he used them with some extravagance. In Appian's words, Ptolemy Philadelphos was 'the cleverest of the kings in raising money, the most splendid in spending, and the grandest in construction'.[1] And one of the main elements in his power, something on which he spent lavishly, was his navy.

Athenaios lists Philadelphos' ships as follows:[2]

2 Thirties	1 Twenty
4 Thirteens	2 Twelves
14 Elevens	30 Nines
36 Sevens	5 Sixes

17 quinqueremes = total 111 ships;
double that total (111 x 2 = 222) in quadriremes, triremes and *trihemioliai*

4,000 other ships which had been 'sent to the islands and the other cities he ruled and Libya'.

In this list Athenaios is specifically itemizing only the big ships, though he was clearly dazzled by their numbers, and his final two categories require some explanation. Ptolemy himself was also clearly entranced by the big ships, but of the ninety-four listed which are larger than quinqueremes, only seventy-one would be useful in a battle – the sixes, sevens, and nines. Those bigger than the nines were of no practical naval use (and the twenty and the thirties were probably almost unmovable) – or at least there is no indication that they were ever used in battle, which is perhaps not quite the same thing. The fleet as we can see it here, therefore, bears all the hallmarks of grandiosity – the building of big ships as prestige items – and it could be that perhaps Ptolemy neglected the more battlefield-useful categories of quinqueremes and smaller.

It seems, however, highly unlikely that Ptolemy had no more than seventeen quinqueremes. Some more are probably hidden in the final category, ships which had been sent out of Egypt to the empire ('the islands', 'the cities he ruled', 'Libya'), which seems to comprise all sorts of ships. This may be the fault of whatever source Athenaios was using. He does not specify what it was, which he normally does when quoting an author. This leads to the theory that he was possibly quoting from an official Ptolemaic government document. Athenaios lived at Rome, but came originally from Naukratis in the Delta, an old Greek settlement. It is quite possible that he consulted old Ptolemaic records which carefully listed the fleet as it stood in Egypt, but only summarized that part of the fleet which was stationed overseas, and so failed to categorize these last groups in detail – hence the supposition that there were some quinqueremes hidden in those categories. It would, of course, be ships of the quinquereme and quadrireme classes which were most likely to be used in cruising about showing the flag, attending to mundane matters, and suppressing piracy.

The last but one category is clearly a summary, claiming that the ships between the quadrireme and *trihemiolia* sizes numbered double the total of the biggest ships. This classification actually would include at least three types, for between quadriremes and *trihemioliai* would have been included the triremes, which are not otherwise listed, but of which some were surely in the fleet. (The *trihemiolia* was a version of the trireme with two full banks of oars, plus one bank with half the full number.) This interpretation implies that the main naval strength of Ptolemaic Egypt lay in its triremes and quadriremes, and this seems unlikely. So, to the big ships Athenaios listed, we need to add the quinqueremes he omitted and the quadriremes, triremes, and other vessels he bundled together in that last but one category.

The final category is infuriatingly vague. '4000' ships may mean 4000 small rowboats, or 4000 quinqueremes. Somebody in the bureaucracy – a huge and immobile category of government employees in Ptolemaic Egypt – perhaps collated a lot of reports and added up a lot of figures. It is possible, however, to link these figures with a passage in Appian's history, in which he lists 2000 barges, 1500 warships from *hemioliai* to quinqueremes, and 800 cabin vessels for naval reviews. There is clearly here an overlap between Athenaios' 4000 ships outside Egypt and Appian's 4300 ships from quinqueremes to cabin vessels. Appian implies that Philadelphos had far more than seventeen quinqueremes, though the precise distribution of types in his 1500 group is unknown – but he does say warships, of which Athenaios had counted only 222 in the classes from *hemioliai* to quadriremes. We may conclude that Ptolemy had a large force of quinqueremes.

The three areas which are said to be where ships were stationed can be identified rather more precisely than those ships. 'The islands' are those in the Aegean, specifically the Kyklades and some others which had been organised into the Island League; a *nesiarchos* was appointed by Ptolemy. 'The cities' were several of the old Greek cities along the western and southern coasts of Asia Minor. 'Libya' is that part of North Africa adjacent to Egypt, also called Cyrenaica. Collectively the phrase could be the bureaucratic term for the Ptolemaic overseas empire, contrasting it therefore with Egypt itself. The empire and the ships go together; they were dependent on each other. The Ptolemaic state was an empire linked by its sea communications: it was a maritime empire.

Two lands which were the most vital parts of the Ptolemaic empire are not mentioned in the list, though it is just conceivable that they are hidden among 'the islands' and 'the other cities'. These were Cyprus and Koile Syria.

Ptolemy I had made repeated attempts to secure control of Syria. In the end, in 301, in the aftermath of the defeat and death of Antigonos I, he had succeeded, later rounding off the conquest with the occupation of Tyre and Sidon. His purpose was transparent to every man of the time: to block any attempt to attack him in Egypt. Twice in fifteen years a major army had penetrated the desert frontier of Sinai to attack him, in 321 and 306; earlier Alexander and his Persian royal enemy had carried out the same invasion in 331 and 343 respectively. Securing control of Syria was therefore Ptolemy's prime foreign policy objective. Having seized it, he and his successors fought repeated wars to keep it – and later, having lost it, they fought still more wars to regain it. Even at the very end, when Rome controlled everything, Kleopatra VII worked hard to extract parts of Syria from the Roman grasp.[3]

The second major Ptolemaic objective was to secure control of Cyprus, in this case partly as a forward defence for Egypt, but also as a naval and military base from which to attack rivals; a subsidiary aim was to gain control of the island's rich resources of timber, metal, and a taxable population. The occupation of Syria blocked the land approach to Egypt from the east across Sinai; control of Cyprus (and the Syrian coast) prevented any naval attack. Ptolemy early on aimed to secure control of Cyrenaica, the group of Greek cities which had been part of the old Persian satrapy of Egypt. This was a further defensive measure. It turned out to be fairly easy to achieve control there, but difficult to maintain it, for the cities were not happy at their subjection; Ptolemy I's first viceroy, Ophellas, became effectively independent, a situation Ptolemy I tolerated since Ophellas was no threat. Later Ptolemy II put his half-brother Magas in the office of viceroy, only to find that Magas was even more determined on independence than Ophellas, and made Cyrenaica into an independent kingdom for a whole generation. He also attempted an invasion of Egypt, but was easily deflected from it when a rebellion broke out behind him.

Magas' kingdom, protected by an alliance with the Seleukid King Antiochos I, could be endured, as could Ophellas' defiance. But Syria and Cyprus were vital defences for the core of the state in Egypt. Yet these forward defences soon ceased to be mere defences and became part of the state, and so they also needed to be defended. Beyond Cyprus, it was necessary to dominate the seaways of southern Asia Minor and the Aegean.

One of the more successful political initiatives of Antigonos I had been to form the islands of the Kyklades into the Island League, with its headquarters on the sacred island of Delos. When Demetrios vanished into the Anatolian interior in 286, this Island League became part of the loot, and Ptolemy was the successful bidder. He had already used Andros (one of the Kyklades) as a base from which to assist Athens against Demetrios in 288–287, which implies that the 'great fleet' he used at that time was powerful enough to deter any attack by Demetrios. Ptolemy was therefore in a good position to influence the Island League when Demetrios was visibly failing. By 287 he had supplanted Demetrios as the League's patron and protector, and was able to back up his influence with supplies of food and other assistance.[4] From these islands, with their many harbours and their recruitable seafaring populations, Ptolemy's fleet could dominate the Aegean; in 280 Philokles of Sidon, still his commander-in-chief in the region, was able to influence the Islanders, and Samos as well. He will have been familiar with the islands from his time as Demetrios' commander; he will also have been familiar to the Islanders themselves.[5] The transition from Demetrios' patronage to that of Ptolemy was clearly an easy one.

Bases in the islands allowed Ptolemy's fleet to reach all the Aegean Sea's surrounding shores, and several of the cities of the Asian coast came under Ptolemy II's control. The reasons were various: some sought protection against an avaricious neighbour, some were spoils in a war, some even came in voluntarily (like the Islanders), others were taken under control to deny them to an enemy. Of the islands near the Asian coast Samos and Kos were Ptolemaic naval bases, and the alliance with Rhodes continued even after Demetrios' downfall. Several Karian and Lykian coastal cities – Xanthos, Halikarnassos,

Amyzon, Miletos, and others – became part of his empire. And, as is the way of empires, Ptolemy's control and influence tended to spread inland from the coast. He placed a relative, Ptolemy son of Lysimachos, as ruler in the city of Telmessos in Lykia. These acquisitions were in addition to the two cities of Pamphylia – Korakesion and Phaselis – which Ptolemy I had seized in his cruise in 309.[6]

The northern Aegean does not seem to have been included. Ephesos, for example, did not fall to Ptolemy II's rule until 262, and then only briefly. Penetrating further north he would have reached into the areas dominated by Antigonos from Macedon, or by Antiochos I, the Seleukid king, who was dominant at the Hellespont. This Ptolemaic restraint, if that is what it was, prevented immediate conflict, but this division of the sea was an obvious source of friction.

The Aegean was one of the cockpits of the Mediterranean world, a source of ships, taxes, recruits, skills, and influence, so it is not surprising that wars were frequent there, as they had been in earlier periods. Since Ptolemy was in control of much of the area, every war involved him; and in each war he found it necessary to extend his control over new territories, if only to protect those he already had. He was also in conflict with the Seleukid kings over Syria, and since Seleukid territories extended into Asia Minor, the Aegean lands were involved. The Seleukids did not have much of a navy until the end of the third century, so were no threat to Ptolemy's fleet. This confined the conflict to land warfare, particularly along the coasts. But this also meant that Ptolemy needed to be dominant on land as well as at sea, so to the need for a powerful navy was added the need for an effective army. Such is the price of Great Power status.

The exact extent of the empire therefore fluctuated, with cities being lost, gained, and regained, and new bases being acquired, but for decades these fluctuations were not seriously damaging to the empire, except at the fringes. Philokles was the first of Ptolemy's viceroys in the Aegean area, active until at least 280. In the 270s the post was held by Kallikrates of Samos, who remained active in the area when his successor was in command.[7] By the early 260s the post was held by Patroklos, a Macedonian. He commanded during one of the wars, the Khremonidean, in which Athens, allied with Ptolemy, attempted to rid itself of Macedonian domination. Ptolemy's main interest, of course, was in restricting the power of Antigonos Gonatas, and freeing Athens was largely a means to that end. Patroklos' policy, then, was as much aimed at preoccupying Antigonos and degrading his power as it was freeing Athens. He set up naval bases at three points around Attika – the island of Gaiduronis (now Patroklu, named after him), Korone, and Rhamnos – from which he could control the seaward approaches. Across the Argolid Gulf to the south of Athens he seized Methana on its defensible peninsula. None of this had a serious effect on the conduct of the war, other than to prolong it.[8]

In the wider Aegean Patroklos extended Ptolemaic control in order to have secure access to these posts. The island of Thera became a Ptolemaic naval base, and another was established on the eastern end of Crete, at Itanos. In such places it was normal for a formal treaty to be agreed between the king's viceroy and the inhabitants, in which the usual guarantees about local autonomy were given. For a small place such as Itanos, of course, the Ptolemaic presence was economically very helpful, providing employment, a market, possible subsidies, and protection (even if some of these places did thereby also become targets).[9] The end of the Khremonidean War in 261, whereby Antigonos fastened a firmer grip on Athens, necessitated the evacuation of the Ptolemaic bases on the Attikan mainland, and the nearby islands, but Itanos, Thera, and Methana remained, as did Ptolemaic domination of the Island League. For Ptolemy II, technical defeat in the war only reinforced his power in the Aegean as a whole.

Places such as Thera, Itanos and the others were not just given a treaty and used as the occasional base for ships. Some were garrisoned, as Thera was, and probably Itanos. Methana, fairly close to the Antigonid forces in Corinth and Peiraios, will have been garrisoned as well. In addition, considerable work of construction was done. The garrisons needed fortified positions, which had to be built, and possibly places for the soldiers to live in. And the ports needed to be made usable by a fleet, not just by a few merchant ships.

Take the two places in Pamphylia which had been taken over by Ptolemy I and remained in Ptolemaic control during his reign and that of his son. Korakesion (modern Alanya) is a huge rock, directly on the coast. The town is situated at the foot of the rock, but it is the rock itself which makes the site valuable from a naval and military point of view. It protrudes into the sea, forming a bay on either side. The eastern bay is now the harbour of the modern town, and has been since the Middle Ages at least, for there are towers and shipsheds at the foot of the rock.

But on top of the rock is a castle, a place large enough to have acted as the capital of a medieval kingdom, with storerooms, a palace, and a mosque. The existing walls are of the Seljuk period, but they were built on the same lines as the preceding Byzantine and Hellenistic walls, which can be seen in certain places, so the summit of the rock was fortified when the place was under Ptolemaic control. The town's existence is recorded earlier, but no previous fortification of the place is known. Alexander did not bother to visit it, though he went to every other Pamphylian city. The climb to the summit is steep and arduous, so it is likely that any civic settlement was always at the foot of the rock; only when Ptolemy I seized the place did the summit become a fort. There are still sections of the walls of the castle which are clearly Hellenistic work, notably at the main gate, and it may well be that the basic harbour works, including the shipsheds, are also based on Ptolemaic work. The work was done successfully enough that the city and its castle were not attacked until Antiochos

III's campaign along the coast in 197. He took a month to capture it, even with overwhelming naval and military force at his command.

Phaselis is a Greek city on the coast of Pamphylia or Lykia (the boundary is not clear), directly across the Sea of Pamphylia from Korakesion. It was a local naval power – its coins show a galley – and Egypt was one of its main trading partners. In the Ptolemaic period it was probably one of the sources of timber for Ptolemy's ships. Like Korakesion, but on a more human scale, the city is built on and close to a hilly peninsula. This gives the city two harbours, north and south, both of which have gentle sandy beaches on which ships could be drawn up. In addition there was a third, smaller harbour, part of the northern port, close in to the city. Much of Phaselis' harbour works have been dated to the third century BC, which means they were constructed during the Ptolemaic occupation, which seems to have lasted until about the middle of the century. The work included a quay beside the small harbour, breakwaters for both north and south harbours, and buildings which were perhaps warehouses.[10]

These two cities, therefore, which were in Ptolemaic hands at Korakesion for a little over a century (309–197 BC) and at Phaselis for something over half a century (309–c.250 BC), attracted a comparatively large investment in new construction by the Ptolemaic government. The actual cost is impossible to calculate, but was clearly considerable. This sort of investment has to be envisaged at the other Ptolemaic ports and posts as well. When Athenaios remarked on Ptolemy II's predilection for spending, he was exactly right. Not only did he maintain the biggest navy in the Mediterranean, and build a whole series of big ships, and be able to deploy an army of substantial size, he had to invest extensively in the infrastructures needed to back up this force – castles, city walls, harbours, barracks, and even whole new cities, as at Ptolemais-Ake in Palestine. But then, empire is always expensive.

Empire also had to be maintained by active campaigning, and by diplomacy and intelligence gathering. In the third century BC the Ptolemaic kingdom was the most powerful state in the Mediterranean region. As such every problem in the region became its problem. We have three cases of this. The western powers, Rome and Carthage, both had diplomatic-cum-naval relations with Ptolemy II. Victories of Rome in Italy provoked him in 273 to establish contact with that city.[11] This was at a difficult time. A war with Antiochos I had begun and Magas in Cyrenaica was threatening an invasion of Egypt. Ptolemy had been friendly with King Pyrrhos of Epeiros, whom Rome and Carthage had just defeated, with some difficulty. Making friends with the victors made sense. At much the same time a Ptolemaic squadron under the command of Timosthenes of Rhodes went through the western Mediterranean. Carthage and Ptolemy were both neighbours of Magas of Cyrenaica, whose ambitions would provide a good pretext for contacts. (Ophellas of Cyrenaica had invaded Carthaginian territory thirty years before; if Magas failed to conquer Egypt, he might turn west.) Timosthenes' ships seem to have sailed as far as Spain, as a result of which he produced detailed accounts of the ports around the Mediterranean – it was, it seems, an intelligence gathering expedition as well as a display of power.[12] These diplomatic initiatives were more or less simultaneous, and were no doubt designed to pre-empt any problems emanating from the west. In fact they stood Ptolemy in good stead later when conflict broke out between the western powers; maybe this was already seen as likely at Ptolemy's court in the 270s; certainly Pyrrhos thought so, and he was in contact with Ptolemy.

A third area of Ptolemy II's diplomatic interests which was demonstrated by his naval prowess was the Bosporan kingdom in the Crimea. The evidence here is a fresco, from Nymphaion, one of the cities of the kingdom on the Strait of Kertch connecting the Sea of Azov with the Black Sea. The fresco shows many ships and animals and the painting of a ship called *Isis*. It is interpreted as a ship of Ptolemy II's fleet, partly because of the ship's name (there is no evidence of Isis worship in the region until considerably later), partly because of the date of the painting (the mid-third century BC, approximately), and partly because of a number of Phoenician characteristics which have been detected in the ship. It is also, rather less certainly, interpreted as one of Ptolemy's quinqueremes. The ship is shown with three banks of oars, and so at first it was interpreted as a trireme, though with fewer oars than expected, but its heavy superstructure implies a quinquereme. Assuming that this was a ship of Ptolemy II, its mission was, like that of Timosthenes in the western Mediterranean, no doubt to impress the local rulers with Ptolemy's power, and it presumably also went to other cities and kingdoms around the Black Sea area. Its captain's mission was also surely to investigate the local powers and their capabilities.[13]

There was another frontier which Ptolemy II exploited. His southern boundary abutted against the kingdom centred at Meroe further up the Nile, in the modern Sudan. The frontier of the two states moved back and forth along a section of the Nile between the first and second cataracts, but Ptolemy II's power ensured that he controlled it. Another part of Egypt's southern frontier was the Red Sea, giving access, in Ptolemy's time, mainly to Ethiopia and Somalia and Yemen, but also, less directly, to lands around the western Indian Ocean. One of the elements of contemporary military power which Egypt lacked was elephants; expeditions went to Eritrea on elephant hunts, but gold and spices were also sought and acquired. These were large and expensive expeditions, and involved the development of port facilities at several places along the African Red Sea coast, which were usually named for members of the royal family – Berenike, Philotera, Ptolemais. This policy also required the renovation of the old canal connecting the Nile near its delta with the Red Sea at the Gulf of Suez.

This was clearly a task for the navy, just as Timosthenes' voyage to the west and the mission of *Isis* into the Black Sea were naval tasks. The elephant hunts were largely military, but there were diplomatic aspects as well. Contact in some way was made with India, even as early as the 270s, and the developing kingdoms in the Yemen and on the Ethiopian highlands were contacted – as was, no doubt, Meroe, which was a source of gold and possibly of elephants. All this was one more element of the great extension of Ptolemaic power and influence, in large part based on the great naval strength Philadelphos had built up.[14]

Intruding as it did into every corner of the Mediterranean and the Red Sea, the Ptolemaic empire was almost constantly at war, and was clearly vulnerable to hostile alliances. Ptolemy II survived the war against Antiochos I and Magas in 274–271 (the 'First Syrian War'), but it had been a dangerous moment. Magas became effectively independent as a result and Antiochos I had managed to invade Syria. The Khremonidean War (267–261) was a technical defeat for Ptolemy, in that Antigonos Gonatas starved Athens into surrender and imposed garrisons and governors in the city, which had been inspired to rise against him by Ptolemy; during the fighting Ptolemy did little to help the city. Yet at the same time Ptolemy's naval strength was now at its greatest, and his grip on the southern and central Aegean was even firmer than before.

The experience of the war – of which we know little – seems to have convinced Antigonos that he needed to exert his own strength at sea. Since becoming king in Macedon he had ruled a state which had required a long period of recuperation after the destruction of people and property by the Galatians. The kingdom therefore required peace, and that is what he had provided. Antigonos' involvement in the Khremonidean War was carefully limited, and the main fighting took place in and around Athens, where he could use mercenaries. By the end of that war, however, a new generation was growing to manhood in Macedon, and the kingdom's armed strength was once more formidable.

There is no indication of Antigonos' activity at sea in that war, but he had control of several of the main shipbuilding cities of the region – Peiraios, Demetrias, Corinth – and given the finance he could have ships built. He had also, of course, saved some of his father's ships from the wreckage in the 280s, though how many he still had by the 260s is not known. These ships, built in the 280s or before, will have been mainly obsolete by the 260s, and may well have required replacing, though some of the big ships probably lasted well. Careful maintenance could well have preserved considerable naval strength, but not employing it may well have been one of Antigonos' economy and recovery measures. He did possess a nine, which helped produce a victory for him, and which he renamed *Isthmia*, probably because it had been built at Corinth; it was one of his father's ships and so had been built fifty years before it was finally dedicated to Apollo at Delos.[15] By the mid-260s he was active at sea once more. After all, Ptolemy's main strength in the Aegean was naval; to contest him, a navy was clearly required.

Only a year after the peace was agreed at Athens in 261 to end the Khremonidean War, a new Syrian War began. Antiochos I died in that year, and his son was not constrained by the peace treaty as Antiochos I had been since 271. And the lesson of fighting Ptolemy separately had been learned. Antigonos was apparently still technically at war with Ptolemy in the Aegean, and Antiochos II was determined to start a new war with him for Syria. They made an alliance – all the more readily in that Antiochos II was the son of Stratonike, Antigonos' sister.

One of the few events we know of in this war is a naval battle near Kos, one of Ptolemy's bases, in which Antigonos' fleet defeated Ptolemy's. The date of the battle is a notorious chronological problem, but between 261 and 255 seems to cover most suggestions.[16] The political and military situation which led up to it is as obscure as its date, but it had probably something to do with a confused series of events in and around Ephesos between 262 and 259. The city ended up in Antiochos' control. Samos also was fought over at the time, though it seems to have remained in Ptolemy's hands in the end. The island of Kos is a little south of this area, and had been a Ptolemaic base for half a century. The fact that Antigonos' navy could penetrate into the Ptolemaic sphere as far as Kos and win a battle there suggests that Ptolemy's attention, and much of his naval power, was mainly directed at the war in Syria at the time. In other words, Antigonos' victory was won over only a part of the Ptolemaic fleet.

One result of this battle, and of Ptolemy's necessary concentration on the war in Syria, was that Ptolemaic power in the Island League faded away in the early 250s, though it may have begun to do so earlier. It was, however, not Antigonos' influence which replaced Ptolemy's, but that of Rhodes. In 258 or a little later, Rhodes's fleet fought that of Ptolemy off Ephesos (part of the confused events mentioned in the previous paragraph).[17] The occasion for Rhodes breaking its old alliance with Ptolemy is unknown, but Rhodes was busy exerting its influence among the islands of the Island League soon after Ptolemy's own influence waned, so it looks like Rhodes was taking advantage of the preoccupation of Ptolemy with the fighting in Syria to join with Antiochos II and Antigonos in a grand alliance aimed at bringing down the Ptolemaic empire. No doubt it had become far too overbearing for Rhodes's comfort.[18]

Ptolemy's fleet was commanded by the exiled Athenian politician Khremonides, and faced a Rhodian fleet. For once we know some details of the fight, from Polyainos. (The date of this battle is as uncertain as are those of Kos and Andros; 258 is the date most generally accepted, but this is by no means certain.) Khremonides brought the Ptolemaic fleet out of Ephesos harbour – and so the city was under Ptolemaic control at the time – and was then confronted by a Rhodian fleet, commanded by Agathostratos. The Rhodians were arrayed in single file, apparently prepared for battle, but they backed off when Agathostratos saw the Ptolemaic formation. Khremonides assumed that his enemy's retreat meant his own victory and put back into port to disembark. Agathostratos then reversed his course, changed formation by thickening up his flanks, so strengthening his most vulnerable area and at the same time narrowing his front, which let him get swiftly into the harbour. He caught the Ptolemaic fleet in the middle of disembarkation and won the victory.

The strength of neither side in the fight is known, nor are the Ptolemaic casualties, in ships or men. But Samos and Ephesos both remained under Ptolemy's control, so it is unlikely that the Rhodians were able to exploit their victory, other than by reinforcing their influence amongst the islands. Nor, so far as can be seen, had the defeat at Kos deprived Ptolemy of any islands: Samos, Kos, Itanos, Thera, and Methana remained his, even if his influence in the Island League faded away. One result of all this was that Antigonos felt secure enough to relax his grip on Athens in 256 or 255, implying that Ptolemy was no longer an imminent nor perhaps an active enemy. Peace was made between Antiochos II and Ptolemy II in 253, and in that year Antigonos became embroiled in a fight with his nephew, Alexander son of Krateros, in and around Corinth and Chalkis; a period of peace for Ptolemy then ensued.

It is, on the face of it, surprising that two naval defeats for an empire built on naval power should have had so little effect, other than a reduction of Ptolemy's influence in the Island League. The explanation may be that the defeats were less serious than the victors supposed or proclaimed. It would be normal for minor victories over the greatest power to be exaggerated. The battles probably involved only detachments of the Ptolemaic naval strength, but the full naval power of its enemies. In each case Ptolemy's real power was only marginally affected, and his enemies no doubt scuttled back to safety as soon as it became clear that Ptolemaic reinforcements were expected.

So, two wars which between them had lasted fourteen years (267–261 and 260–253) hardly dented the Ptolemaic position in the Aegean. After peace arrived, from 253, Ptolemaic influence increased once more in the islands, though no new *nesiarchos* was appointed, so far as can be seen. Antigonos had dedicated his victorious flagship, *Isthmia*, at Delos after the battle of Kos, and in 253 he instituted two new vase festivals there, named for himself and his sister, *Antigoneia* and *Stratonikeia* (Stratonike died that year); Ptolemy replied in 249 and 246 with festivals called *Ptolemaieai*.[19] By the time of his death in 246, Ptolemy II Philadelphos had regained the influence in peacetime which he had lost in war.

Almost as soon as his son and successor, Ptolemy III Euergetes, became king, he was involved in yet another Syrian War with the Seleukid king. Antiochos II had also died in 246, a little after Ptolemy II, and the Seleukid throne was then disputed between his eldest son, Seleukos II, and the infant son of his second wife. Then, when the child and his mother had been killed, Seleukos faced a rebellion by his brother Antiochos Hierax. Antiochos II's second wife had been

Berenike, the sister of Ptolemy III. Ptolemy claimed to be attempting to support her, then to rescue her, and finally to avenge her.

He did so by means of a naval expedition to the Syrian and Kilikian coast, using more than one detachment of ships.[20] His base at Cyprus would be one source of the expedition, but Ptolemais-Ake in Palestine was probably another. He had no difficulty with such an expedition, since Seleukid naval strength in the area was negligible, and the Seleukid military defences were clearly paralysed by the internal political crisis. For a time he held control of North Syria at least as far as the Euphrates River, though trouble in Egypt soon called him home. Meanwhile another of his father's foreign policies had brought trouble in the Aegean.

Antigonos Gonatas had lost control of Corinth to his nephew Alexander, who died about 247; the city was inherited by Nikaia, Alexander's widow. Both Antigonos and the leader of the neighbouring Akhaian League, Aratos, aimed to gain the city from her. Aratos had already struck an alliance with Ptolemy II. Antigonos succeeded by sheer effrontery in seizing Corinth (he climbed up, knocked on the gate, and went in when it was opened). A Ptolemaic fleet came up to the waters off Andros Island, where its position cut Antigonos' communications between Macedon and Euboia to the north and the Macedonian holdings of Peiraios and Corinth. When Antigonos brought up his own fleet, perhaps to test the resolution of the Ptolemaic commander, a battle resulted; like Kos, it was Antigonos' victory, and like Kos, its results were a good deal less than its renown.[21]

The Ptolemaic revival in the islands reversed, and Ptolemy III's influence in the Island League was negligible. It was not, though, replaced by Macedonian influence, still less hegemony, nor even by that of Rhodes.[22] It seems that Antigonos established his influence on Keos, and perhaps on Andros, to protect his access to Peiraios and Corinth – a reasonable precaution after the Ptolemaic attempt to interfere, and bearing in mind the continuing Ptolemaic control of Methana. But two years later, in 243, Aratos, without much help from Ptolemy, seized Corinth from Antigonos' control, this time definitively. This acquisition established the Akhaian League as a major political player in Greece, and as an opponent of Macedonian influence. Antigonos still held on to Peiraios, however.

Neither Ptolemy II nor his son were ever really very interested in extending their power over the Greek mainland, though influence was always welcome, and an alliance with Aratos and the Akhaian League was sufficient. The only post they maintained on the mainland was Methana. The end of Ptolemaic influence in the Kyklades after 245 was of minor importance, for Itanos and Thera continued to be held, as were several islands off the Anatolian coast, notably Samos and Kos. What Ptolemy was really intent on was domination of that coast, both in the islands and in the cities, for there his enemy was always the Seleukid kingdom, which controlled the Anatolian interior. So in 245 the defeat at Andros was largely ignored, and whatever influence was lost in the islands was more than made up for by the extension of Ptolemy's influence and lands along the Asian coast. Once again, as with the fights at Kos and Ephesos, a naval defeat had probably been inflicted on only a detachment of the main Ptolemaic fleet. We know that a substantial part of that fleet was active in Syrian waters at the time, so the Aegean force was less than the total potential Ptolemaic fleet.

The collapse of the Seleukid state in the 240s provided Ptolemy with an unparalleled opportunity to pick up as many pieces – cities, outposts, islands, allies – as he wished. His choice reflected the maritime basis of his imperial power. In Syria he held on only to Seleukeia-in-Pieria, the main Seleukid port city, leaving only a single Syrian port, Laodikeia-ad-Mare, to Seleukos II, a city which is in an awkward geographical position for communicating with the interior, being backed by a mountain range. Several small cities on the Kilikian coast were kept, and the old posts in Lykia and Karia were maintained (though Phaselis was let go, it seems). Ephesos was, of course, recovered and held at the peace, and Lesbos was added to Kos and Samos as islands dominating the nearby coasts. The Hellespont came into Ptolemy's control by his acquisition of Lysimacheia and the Thracian Chersonese; two old Greek cities on the Thracian coast, Ainos and Maroneia, were also his.[23]

The result was a reorientation of Ptolemaic Aegean control, from a domination emanating from the Kyklades, to a long line of coastal and island bases along the length of the Asian coast from Kilikia to Thrace. Samos and Ephesos were major bases, but it appears that the headquarters of the whole Ptolemaic position in the Aegean was now at Thera, well fortified and garrisoned, with a very useful and well-protected harbour, and well distant from any of Ptolemy's enemies. From 240 this was to remain more or less the position for the next forty years, give or take a city or so.

The Battle of Andros, therefore, had perhaps even less in the way of political results than merely removing Ptolemaic influence from the Island League. The subsequent failure to revive that influence suggests that it was a deliberate decision by the Ptolemaic government to abandon the League, perhaps because of its proximity to the Antigonid communication route between Peiraios and Macedon which would obviously cause continual friction. It seems clear that Antigonos and Macedon were relatively minor irritants to Ptolemaic power, when compared to the major threat from the Seleukids. And it was that threat which was clearly being watched and guarded against by the disposition of the naval bases as they were re-organized in the late 240s.

Chapter 7

Agathokles and Carthage

The eastern and western theatres of war in the Mediterranean only rarely intersected with each other. In the west the strategic prize was always Sicily, and in particular the great city of Syracuse, so one of these intersections had been the colonization of that island and southern Italy from Greece; another was when Athens attempted to seize Syracuse as a precursor to the conquest of all the west. Economic relations between the two regions were constant, particularly the export of grain to hungry Greece, but politically contacts tended to be less than continuous.

There were five cities in the west which had significant fleets of warships, though most other coastal cities had a few ships which were mainly used for harbour defence and combating pirates. Syracuse was one of these maritime powers, as was Carthage in North Africa; in southern Italy Tarentum was the most significant naval city. These three were the main contenders for command of the seas around Sicily; outside the Sicilian area the only worthwhile maritime powers were Rome and Massalia in southern Gaul, though the Etruscan cities were active at sea in an informal, privateering or piratical role. Massalia was well distant from the rest, but Rome became steadily more involved in affairs in southern Italy and Sicily from the 320s.

Previous bruising encounters with the Greeks in Sicily had pushed the several Phoenician colonies into a close political relationship led by Carthage, and this is frequently seen as a Carthaginian hegemony, or even empire. In practice it was a fairly loose political arrangement. Carthage itself was one of the major cities of the Mediterranean world, and in the fourth century BC it had began to make serious efforts to conquer and develop its immediate hinterland (the modern Tunisia), partly to reduce the threat from hostile locals, and partly to ensure an agricultural supply base for the city. But the city was especially active commercially, and the trade routes along the North African coast were under its control. A series of trading posts dotted that coast as far as the Atlantic coast of Morocco, and there were more of them along the opposite Spanish coast, and in the Balearic Islands and Sardinia. Many of these settlements had independent origins, and perhaps were less than enthusiastic about Carthaginian control.

Other cities in western Sicily and southern Sardinia were more amenable since they were uncomfortably close to the Greeks of Sicily. The great Carthaginian fortress of Lilybaion on the western tip of Sicily was a major centre of military power.

Syracuse had been ruled by the great tyrant Dionysios I until his death in 367, and then by his son. They had controlled all Greek Sicily as, in effect, a kingdom, and had extended their power over southern Italy (Magna Graecia to the Romans) and into the Adriatic and Tyrrhenian Seas. Dionysios I was later said to have commanded a fleet of 400 ships – or 500, the figures vary, and are probably exaggerated – and had been the inventor of, or the inspiration behind the development of, quadriremes, quinqueremes, and even perhaps a six.[1] His kingdom progressively collapsed under his son, Dionysios II, and it seems that Syracusan naval power decayed as well. The third centre of power, Tarentum, was a good deal weaker than either Carthage or Syracuse, but was sufficiently distant from both to be able to maintain a local control in the heel and instep of Italy, and in the local seas.[2]

The three cities largely owed their prominent maritime power to the fact that they had been founded at sites with sheltered and capacious harbours. These had been improved and extended by excavation and building, and the cities themselves had grown and become heavily fortified. It was the combination of big well-equipped harbours and large well-fortified cities that gave them their power, both economic and naval. Syracuse was built on the northern and western shores of its Great Harbour, though the oldest part of the city was on Ortygia, an island, now a peninsula, which sheltered the harbour from the open sea, and that harbour had shipsheds, built by Dionysios I to accommodate 300 vessels. The heights of Epipolai to the north provided a fortifiable area to form a substantial defence from attack by land, or a tyrant's base of power, depending on the political situation. Tarentum had a very similar situation, with a wide bay, now the Mar Grande, partly delimited from the sea by a pair of islands, and a smaller bay, the Mar Piccolo, sheltered by the peninsula on which the city was built. A wide area was walled, and a powerful acropolis dominated the whole. Carthage had the least welcoming maritime situation, but it lay at the head of a wide bay, on a peninsula which had been powerfully walled. Two lakes, to north and south, formed the peninsula; they were shallow but provided a useful defensive barrier. The shore provided a comfortable beach to draw ships up on, but the city had excavated two artificial harbours to improve its resources. There was a square commercial port, and a circular naval port. The former had warehouses around it, the latter shipsheds. All three cities were thus deeply committed to the sea and sea power, but it was still political will which was needed to, so to speak, launch the ships.

In the 340s both Syracuse and Tarentum (which had been in Dionysios' sphere of control for a time, but recovered its independence after his death) faced major difficulties. Tarentum felt itself to be under threat from the Italian peoples inland; control of Syracuse was disputed between rival tyrants, and so the city had lost its dominant position in Sicily. Both cities turned in their desperation to their old Greek city-founders for help, Tarentum to Sparta, Syracuse to Corinth.

This reaction to their political problems was to become a habit, particularly at Tarentum, where the employment of Archidamos of Sparta in the late 340s was followed by calling in four more soldier adventurers, from Sparta and Epeiros, during the following sixty years. At Syracuse the expedition of Timoleon of Corinth solved Syracuse's internal problem for the time being, and resulted in the removal of a whole series of tyrants in several other cities of the island, and in the defeat of a Carthaginian attack.[3] But the subsequent division of the island among many independent cities was unstable; and it was not long before a new tyrant, Agathokles, took power in Syracuse, in 317.

Most of the warfare which all this involved was inevitably conducted by land, but once Carthage became involved there was an obvious maritime element to it as well. And not only Carthage. Any commander summoned from Greece to either Syracuse or Tarentum had to reach his employer by sea. Archidamos set out from Sparta with 'a few ships' in 344;[4] Timoleon, who left Corinth for Syracuse at more or less the same time, travelled with ten ships. At the time Carthage was blockading Syracuse with a 'great fleet', which is otherwise stated to be 150 strong, and next year the same fleet is identified as consisting of triremes. Two years later, in 342, after building new ships, Carthage sent a major expedition against Sicily, 200 warships convoying a thousand cargo ships.[5]

It is clear that in the 340s Carthage was the predominant power at sea in the west, but the only uses the city made of its warships were to convoy the expeditionary force to Sicily, and to blockade Syracuse. There is no sign of any Syracusan vessels at sea, though this may be because the city was internally divided between rival warlords, and that the priority was to remove the Carthaginian besieging forces – as well perhaps as the fact that the number of Carthaginian ships was so overwhelming that no credible challenge could be mounted. Both Dionysios I and Dionysios II had controlled major navies (that is, until 344), and it is most unlikely that the ships would have disappeared in only a few years. The city was divided between at least two would-be tyrants; clearly neither could let the other use the ships; there must have been some left out of the 400, or 500, which Dionysios had had twenty years before, but the continuing blockade by Carthaginian ships moored and patrolling near the harbour entrance, does show that no Syracusan vessels could challenge them. It was even possible for the Carthaginian commander to detach 20 ships from the blockading squadron to try to intercept Timoleon's ten vessels from reaching Syracuse. In an early sign of his cunning, Timoleon successfully distracted the Carthaginian captains until he could cross over to the island.[6] Nevertheless it is perhaps instructive that Carthage is said to have added more warships to the fleet of 150, as though the potential sea power of the Syracusans was seen to be threatening.[7]

Timoleon defeated the Carthaginian attack, and successfully made peace soon after, and this settlement held for a generation.[8] So the possible naval threat from Syracuse faded; no Carthaginian fleet as large as the 200 ships of 341/340 is heard of for the next forty years.

In the years after 320, however, another Syracusan tyrant, Agathokles, gained power in the city, and in 313 war began between Syracuse and Carthage. Agathokles had some naval power. The size of his fleet is never precisely stated, but it can be calculated, from the events of the next years. He took twenty ships to sea early in 310, took sixty to Africa later that year, and the next year another twenty sailed. After that he was able to set sail with just seventeen ships. So he had at least a hundred warships at his command at the beginning of the war. It is also clear that for some time Agathokles did not, or could not, use his fleet.

The new war grew out of the conflict within Greek Sicily, in particular between Agathokles and his supporters on the one side and the oligarchic exiles from Syracuse who were in opposition to him on the other. The exiles were able to ally themselves with other cities whose rulers – usually other oligarchs – were opposed to, or scared of, Agathokles. In 315 one of Tarentum's hired saviours, Akrotatos of Sparta, was able to sail directly from Tarentum to Akragas, in southern Sicily, along with a group of those Syracusan exiles in twenty ships, without being harassed or intercepted by any of Agathokles' forces.[9] And when, after Carthaginian efforts to mediate peace had failed and Akrotatos had failed also and was driven out, Agathokles attacked Akragas by land, the city was relieved by the Carthaginians with a fleet of just sixty ships, without any naval threat emanating from Syracuse.[10] It seems probable that Agathokles could not afford the manpower to get his fleet to sea and at the same time field an army.

This is confirmed by the war which followed. Carthage sent an expedition by sea against Syracuse city using only fifty pentekonters, smaller, lighter ships much more vulnerable to attack than triremes. Some of these were later captured by Syracusan forces, but it is obvious that the Carthaginians had no real fear of a major Syracusan naval reply.[11] Next year, 311, Carthage sent another expedition to the island, composed of 130 ships plus freighters; 60 of the warships and 200 of the other ships were lost in a storm, but this had no obvious effect on operations.[12] Again it is clear that the Carthaginians had no fear of a naval riposte by Agathokles. We must conclude that Agathokles was unable to put his ships to sea.

Agathokles' war was naturally conducted mainly on land, since his enemies were in the cities of Sicily, and these could most easily be reached overland. The Carthaginians meanwhile pursued their normal strategy of blockading Syracuse and transporting their armies, both by sea. From events in 310 and

309, however, it is clear that Agathokles had considerable naval power had he chosen to use it. In that year he lost twenty ships, captured by the Carthaginians at the 'straits', and it is probable that other minor sea fights took place.[13] Later that year, he put to sea with sixty warships. They are defined only as ships, but since they were rowed and had rams, it is clear they were war vessels. He was mounting an expedition out of Syracuse to invade Africa. On leaving Syracuse harbour he was pursued by the Carthaginian blockading fleet (a factor which allowed some much needed grain ships to reach Syracuse). When he landed in Africa he burnt his own ships, either to prevent them falling into Carthaginian hands, or to avoid having to guard them. So he had his whole army available in the subsequent fight on land – another indication of his lack of manpower. (The Carthaginians, astonished and pleased at this destruction, landed and collected the metal rams to display their success – hence we know that the ships were the warships).[14] Agathokles had thus lost, or perhaps it would be better to say he had used up, eighty ships in that year. Next year a squadron of twenty Syracusan ships is mentioned, of which ten were lost – the others were saved by men from Megara Hyblaea, where the ships were beached; another set of seventeen is alluded to later, and this possibly included the ten which were saved.[15] At the beginning of 310, therefore, Agathokles had at least 107, possibly 117, warships in Syracuse harbour, of which he lost 90 in the next two years. Presumably 107 ships were not enough to drive off the Carthaginian blockading squadron. By 307 the squadron in Syracuse was down to seventeen ships. Carthage was clearly supreme at sea from the start of the war.

While Agathokles campaigned in Africa, and mounted a land blockade of Carthage city, the Carthaginian army in Sicily was besieging Syracuse as well as blockading it by sea, so the two forces were each besieging the other's main city. At this point it was command of the sea which was the vital factor. Without ships Agathokles' siege of Carthage was impossible to win, since the Carthaginians could always bring in food and reinforcements; with ships the Carthaginian siege of Syracuse would inevitably eventually succeed. The Syracusan dilemma was demonstrated by the unsuccessful attempt to bring in more grain ships. A squadron of twenty triremes (the third of those Syracusan forces mentioned in the last paragraph) slipped out of the Syracuse harbour and sailed up the coast to await the provision ships' arrival. The Carthaginians did not see them go, but noticed that they were missing soon enough; no doubt it was the blockaders' practice to look into the harbour every morning to count the Syracusan ships. Thirty Carthaginian ships went in chase, caught up with the Syracusans, and defeated them. The Syracusans beached the ships in an attempt to escape, but the Carthaginians dragged ten of them back to sea with grappling hooks, though the Greek crews escaped. Meanwhile the grain ships could not enter the harbour without an escort, and the blockade by sea continued. The Carthaginians had therefore had a larger fleet than thirty ships on blockade duty, since they had other ships still at the harbour entrance. The action of the Carthaginians in seizing the ten ships presupposes that the ten they did not capture would be refloated and returned to Syracuse.

Blockading an ancient port did not require the blockaders to be at sea all the time. No doubt one or two ships kept watch for any movement in the port, but the main body was beached. This would preserve the ships, avoid storms, and preserve the health of the crews as well, while the ships could be launched quickly, as the Carthaginians showed in the chase. So a blockade actually consisted of a camp on shore. It was not a tight blockade, for ships could get into harbour by surprise, at night, and by using greater speed than the blockaders. All of these methods were used successfully by Syracusans in this siege.

The loss of ten ships appears to have virtually finished off the Syracusan naval strength, and it allowed the Carthaginians to reduce their blockading force. Both sides clearly knew exactly how many and where their enemies were. Agathokles now deserted his army and returned to Sicily in a ship he had had constructed, showing that he had not captured any Carthaginian ships, warships or otherwise, during his campaign. Back in Syracuse he organized a naval sortie next year, but he had only seventeen ships of his own with which to do so. He hired eighteen Etruscan ships, and these got into the harbour secretly by night. The Carthaginian blockading squadron had been reduced to only thirty ships by this time, presumably in the knowledge that the Syracusan naval strength was about finished – without the Etruscans, Agathokles' naval force was clearly outnumbered. Agathokles took out his own ships, an action which provoked the Carthaginian squadron into chasing him; these were then caught between his two forces as the Etruscans came out behind them. The Carthaginians lost five ships in the fight, and it appears that they then lifted the blockade. Agathokles is now said to have 'ruled the sea', which can only mean that he was able to get in and out of Syracuse harbour without having to run the blockade; it also implies that the Carthaginians were concentrating all their efforts on eliminating his army in Africa, and presumably patrolling to prevent reinforcements arriving. They had certainly not lost so many ships that Agathokles' seventeen vessels 'ruled the sea'.[16]

Agathokles returned to Africa once more, evading any patrols, but was defeated in battle soon after, and his army broke up. He deserted it again, and those who survived were largely bought off by the Carthaginians. Back in Sicily Agathokles faced both the continuation of the Carthaginian war and strong opposition from a coalition of his internal enemies – cities and exiles. The ending of the fighting in Africa meant that Carthage could now reinforce its army in Sicily; it also presumably freed the Carthaginian ships for transport and blockading duties if needed. There was thus a real danger that Agathokles would be overwhelmed by his various enemies, who only had to combine to achieve this. Like any sensible diplomatist he concentrated on dividing those enemies and then removing one of them. Since the exiles and the Greek cities were irreconcilable, it was the Carthaginians, heavily damaged in their African lands by his invasion of Africa, who proved the easier to deal with. By offering to cede the western third of the island to Carthage, Agathokles persuaded them to make peace without too much difficulty. He then crushed the Sicilian opposition. For the next decade and a half Agathokles ruled two thirds of Sicily as king – he took the title in imitation of the contemporary military man in the east – while the Carthaginians controlled the western third of the island.

The war had been a lesson to Agathokles in the application of sea power, and as king he built up his own navy. He was able, only a year after destroying the last remnants of opposition in Sicily, to take a small fleet to force recognition of his overlordship in the Lipari Islands, north of Sicily. The story is related as though it was merely a looting expedition, but it was also clearly a conquest. The islands were in a strategic position between Sicily and Italy, and events in Italy inevitably concerned any Sicilian ruler. He did not, however, have a large fleet, and is said to have lost eleven ships on the voyage back.[17]

Agathokles needed a large fleet as a deterrent to Carthage, for that city's major strategic ploy during the war had been the blockade of Syracuse. Since they divided Sicily between them he and Carthage were always liable to fall into disputes. Agathokles could not allow another blockade to develop, for this would probably result in the collapse of his control over the rest of Sicily, where his enemies had been beaten but remained unconvinced of his right to rule them. A major naval force was also a standing threat to Carthage itself, since it was just the lack of this which had deprived Agathokles of victory in the invasion of Africa. Had he been able to blockade Carthage as well as assault it from the land side the result of the war may well have been different.

In addition Agathokles must have had an eye on developments in Italy, where Rome had been campaigning since the 320s to establish its hegemony, and where Tarentum was still looking for Greek military adventurers to help her. Archidamos of Sparta had been followed by King Alexander of Epeiros, Alexander the Great's brother-in-law, who had been fairly successful for a time. After his death in 330 it was fifteen years before another soldier had to be hired, but since this was Akrotatos, who was diverted to Akragas, the situation in southern Italy was clearly satisfactory, at least from Tarentum's point of view. A new expedition, by Akrotatos' brother Kleonymos, was mounted in 304. The Tarentines sent him money with which he hired mercenaries at Cape Tainaron, and ships to bring him to the city. His power was such that the enemies of Tarentum were quickly cowed, and his naval power allowed him to gain control of Corcyra on the Balkan side of the Ionian Sea, and to mount a raid the length of the Adriatic Sea, which ended in defeat. He soon made himself unwelcome in Tarentum, and was induced to leave.[18]

Amid all this, Kleonymos had appeared to threaten Agathokles, who could scarcely ignore the appearance of Akrotatos' brother at the head of a major army. He moved into the toe of Italy about 300 BC, and about the same time took control of Corcyra.[19] That is, since the Lipari expedition he had built up a considerable naval strength, sufficient to carry an expedition to the Balkans. Yet he kept clear of Tarentum, which suggests a balance of naval power between the two. He did not keep Corcyra, but handed it on to King Pyrrhos of Epeiros as dowry for his daughter Lanassa.[20]

Meanwhile the Romans had been extending their power in central Italy. They collided with Alexander of Epeiros, and with Kleonymos, and Agathokles' employment of Etruscan ships and sailors might be seen by the Romans as a threat, but both Romans and Syracusans were wary of Tarentum. The Romans had begun to develop a naval arm. In 313, when the war in Sicily began, a Roman colony had been sent to the Pontiae Islands, off the Campanian coast.[21] By that time firm alliances had been forged between Rome and Capua, Neapolis, and other Campanian cities, and several of those cities were minor naval powers. In 311, while Carthage was blockading Syracuse, the Romans instituted a new office, annually elected *duumviri navales* – two men to command the ships.[22] Each was to have a squadron of ten ships, and next year one of these was used in an unsuccessful landing in the rear of an enemy city in southern Campania.[23]

The fact that Rome organized its ships into a formal pair of squadrons does not mean that these were the first Roman warships, nor does it mean that these were the only warships Rome had available. Like all the other cities with coastlines Rome had had warships before 311, and as other cities came under Roman control their ships could be deployed in Rome's interest. The duumviral fleet of 20 ships was instituted presumably because Rome's military needs required it; it may have become clear that the other cities were unwilling to use their ships in Rome's interests, and Rome's political system in Italy was not yet so tightly controlled that Rome could insist. It would take a new organization to combine the small Italian city fleets into one large fleet, and the conquest of Italy scarcely required such an organization and a dedicated commander, at least for the present. But the Italian city fleets did exist; the twenty Roman ships were not the only Italian city fleets in existence.

It was not Rome's naval strength which drew attention, however, for around 300 BC this was still small; rather it was the progress of the city in conquering and organizing central Italy. The fighting in Sicily and Africa scarcely left anyone there time to do anything about it, but as soon as the fighting ended both Carthage and Agathokles took diplomatic and naval precautions.

In 306, the year after the peace in Sicily, Carthage made a new treaty with Rome. There had been earlier agreements between the two cities, which defined the areas in which they could operate, largely to Carthage's advantage. These had been successful in that they had prevented conflict, and it was probably just one of a whole series of treaties with other cities with which Carthage had relations. Carthage had to take account of the new situation in Italy, where Rome was now predominant, and Rome had to take note of the new situation in Sicily, where, for the first time since Rome's expansion began, the Greeks were united and powerful. The war with Agathokles had been a very bad experience for Carthage, which seems to have believed that its African territories were automatically out of reach of any enemy, so Agathokles' invasion had been all the more shocking. Now the city's government determined to lay down exact boundaries. The treaty of peace with Agathokles had done that in Sicily, and now the Roman treaty did the same, but more widely. Rome agreed that 'Sicily' was in the Carthaginian sphere, though it was obviously the Carthaginian territories in Sicily which were meant, rather than the whole island. This provision was in addition to the earlier, similar provisions concerning Rome's exclusion from Sardinia, much of Spain, and all Africa except Carthage city.[24]

Also in the treaty, Carthage, which in earlier treaties had been free to take actions anywhere in Italy outside Rome's dominions, now agreed that all Italy was outside its bounds, which obviously meant the Roman parts of Italy, for there were still huge areas of Italy which were both independent and powerful – Tarentum, for example – which Rome could not speak for. But the treaty did give both cities a clear idea of the sphere of influence and future action of the other. In particular Rome was still visibly expanding its Italian territories, and any further Roman conquests would automatically become covered by the treaty. No doubt Carthage reserved for herself the possibility of expanding into Greek Sicily. The kingdom of Agathokles was almost as unstable as a disunited Sicily, and the Carthaginians could well have expected the old divisions to revive at any moment.

At some point, possibly in 320, but more likely after 306, Rome and Tarentum had made an agreement concerning the range of each other's naval activity, whereby Rome agreed that none of her warships should sail east of the Lacinian Cape,[25] which was just west of Kroton (though this does not appear to have prevented Tarentine ships from passing that cape westwards). This is a very similar, if more restrictive, agreement to that with Carthage, a definition of a boundary which is also a delimitation of spheres of influence. But it did not affect the land operations of Rome in the rest of Italy, nor did it affect any naval operations of Tarentum. In 303, Tarentum called in Kleonymos of Sparta, possibly at about the same time the Roman-Tarentine treaty was agreed.

Also relevant is the fact that Kroton, west of the Lacinian Cape, was soon (in 295) captured by Agathokles. He was clearly concerned over what was happening in Italy. In his career before gaining power in Syracuse he had fought as a mercenary commander for Tarentum, and had contacted other Italian cities at various times. He had more than one contact with Etruscans, notably the ships he hired in 308. It is reasonably certain therefore that he will have been well up-to-date with events to his north. He must have been particularly concerned at the contacts between Rome and Carthage, and the terms of their new treaty will have quickly become known to him. (The treaty was later virtually suppressed by the Romans, and our main knowledge of it comes from criticism by Polybios of an account by Philinos of Akragas, a Greek historian who was contemporary with the treaty and Agathokles.) In the face of a possible hostile alliance preparations were needed. A major naval force would clearly be of great service if Rome proved hostile, or if Rome and Carthage, or Rome and Tarentum, or Tarentum and Carthage, or even all three, combined against him. Agathokles therefore began building new ships. His raid on the Lipari Islands in 304 intruded into a city which had already been contacted by Rome, though not incorporated into its system. The islands had also been used as a base by the Carthaginians in the past. Bringing them into his kingdom was a reasonable pre-emptive move to lock out both powers who had so recently pretended to carve up the whole region between them. Agathokles lost eleven of his ships in a storm on the way back, but the fleet was clearly considerably larger than that already.

The new fleet developed by Agathokles after 306 was influenced by the fleets which had become active in the east. Despite the earlier development of bigger types of ships, quadriremes, quinqueremes, a six, in Syracuse, it is very noticeable that the wars in Sicily between 344 and 306 were all fought with triremes, pentekonters, and cargo ships. There is never a mention of anything larger. It must be presumed that the larger ships built by Dionysios I and II before the 340s had become unserviceable. Timoleon did not use ships at all in his wars – except to sail to Sicily – and in the peace which followed his victory and retirement there was no call for warships. And so, not apprehending any threat, neither Syracuse nor Carthage had any call to invest in the bigger ships. One must assume that replacement building of triremes was all that was done.

But, just as Agathokles clearly had good information on the political and military situation in Italy, he was also well-informed about developments in the east. In 299 he used his new fleet in an attack on the forces of Kassandros at Corcyra. His Greek soldiers were especially pleased that they met and beat Kassandros' Macedonians, the conquerors of the world. While there, Agathokles saw to it that the Macedonian ships were destroyed.[26] This action placed Agathokles in a commanding position in the whole Ionian Sea region. Apart from Tarentum, there was now no naval power of any pretensions between Carthage and the Aegean except Agathokles. The destruction of the Macedonian ships was therefore a deliberate act to secure his command of that sea.

Agathokles followed up this stroke by further deliberate measures. In about 295 he seized control of the city of Kroton, and a year later also of Hipponion, and possibly Lokroi as well.[27] This gave him control of Calabria, the whole toe of Italy, and it is very obvious that he had moved into the area between Rome and Tarentum in part to keep these two powers apart. This Calabrian expansion (he was also friendly with Rhegion, the last remaining independent city in Calabria) gave him control of another stretch of the Tyrrhenian coast, to add to the Lipari Islands and his two-thirds portion of Sicily. He was thus a major player in the area, interposing his power between Rome, Carthage, and Tarentum, and his kingdom was now strongly based on a powerful fleet. Had his kingdom endured, the history of the next century would have been very different.

At this same time Agathokles was active diplomatically in the affairs of Greece, taking, as the raid on Corcyra indicates, a position opposed to Kassandros. He contracted a marriage alliance with Pyrrhos, king of Epeiros, probably in 295, the year he took Kroton. At some point also he himself married Theoxena, one of Ptolemy I's daughters; for Ptolemy this neutralized the Sicilian fleet in his conflict with Demetrios; for Agathokles it confirmed his domination of the Ionian Sea. In addition to all this he was undoubtedly well-informed of naval developments. His fleet is said to comprise 200 ships,[28] a total which included quadriremes and sixes, and gave him a rough parity with Ptolemy in ship numbers; a few years later, when Pyrrhos took over Agathokles' navy, it is said to have comprised 120 aphracts and twenty cataphracts, which could have been almost any size; there was also a nine – the 'royal nine' – which was no doubt Agathokles' flagship.[29]

Agathokles is said to have made preparations to invade Africa once more, gathering a fleet of 200 ships, though he died before he could begin. He was 72 years old, and this makes any new invasion of Africa fairly unlikely (though the Macedonians fought each other into their 80s). On his death the kingdom collapsed. None of his family could hold it together; Syracuse overthrew his statues and instituted a democracy; Akragas seized the chance for independence; Messana fell to a group of Italian mercenaries. War followed, and the cities fell into tyrants' hands.

Carthage intervened early in the conflict to ensure that the kingdom broke up. During the wars between the tyrants Hiketas of Syracuse, elated by a victory, raided the Carthaginian part of Sicily, but was then defeated. New tyrants seized power in Syracuse, one holding the city, the other Ortygia. Meanwhile Tarentum called in yet another military adventurer, Pyrrhos of Epeiros, to save it from its Italian enemies, who this time included the Romans. The Sicilian Greeks took note of his successes and soon asked him to do the same for them.

Pyrrhos was therefore at war with both Rome and Carthage at the same time. Carthage had attacked Syracuse in its usual way, by blockading the harbour with a fleet of a hundred ships, and by land with an army reputed to be 50,000 strong.

Syracuse, paralysed politically, still held Agathokles' fleet, if only it could be used. Pyrrhos sailed to Tauromenion first, and then into Syracuse, whence the Carthaginians had prudently withdrawn. He had arrived with a major fleet of his own, and added to it the ships at Syracuse – 140 ships and a nine – making a fleet over 200 strong. The Carthaginian blockading fleet, put at a hundred originally, had been reduced by the detachment of thirty ships to Rhegion with the intention of either blocking Pyrrhos' passage, or distracting him.[30]

Capitalizing on his early popularity in Sicily, Pyrrhos gathered an army and drove the Carthaginians back to Lilybaion, but his two-month siege was unsuccessful. Behind him, Tarentum was annoyed that he had left it to face Rome alone, and the Greeks of Sicily felt that he was becoming a tyrant. No doubt he dreamed of uniting Epeiros, Magna Graecia, and Sicily into a great kingdom, and perhaps he aimed to invade Africa. He was the son-in-law of Agathokles, after all. But the parts of his putative kingdom seceded from beneath him, and he returned first to Italy, and then to Epeiros.

His power and energy had been such that Rome and Carthage had been drawn into an alliance. Already in the spring of 278, while Pyrrhos was still at Tarentum, the Carthaginian fleet of 120 (or 130) ships brought an admiral, Mago, to Ostia with plenipotentiary powers. The Senate was reluctant. It already looked as though Pyrrhos would go to Sicily, and once there he would be Carthage's problem. So Mago made moves to negotiate with Pyrrhos; the Senate swiftly agreed to an alliance.[31] The terms were carefully negotiated to preserve each ally's area of power and influence, and the clause providing for mutual assistance was only invoked once, when Carthaginian ships took Roman soldiers to attempt to capture either Lokroi or Rhegion and to protect Thourioi.[32]

How effective the Carthaginian squadrons were is not clear. One was blockading Syracuse with a hundred ships, from which thirty were detached to Rhegion; Mago had 120 (or 130) at Ostia; chronologically these are all separate, and it could be that we are seeing only one fleet used in several places. It certainly seems unlikely that Carthage disposed of 230 ships at this time, and it would be more economical to see only a single fleet of, say, 120 ships used in successive tasks. When Pyrrhos had his combined fleet of 200 ships under his command, there is no sign of any Carthaginian maritime opposition, which rather implies that Pyrrhos' fleet outnumbered Carthage's.

On the other hand, when he was leaving Sicily, Pyrrhos marched by land as much as possible. He was harassed by the Mamertines of Messana, and he had to fight a Carthaginian squadron which intercepted him as he was crossing the Strait. Presumably he had left the Syracusan ships behind, and was moving by land because he did not have the naval strength to make a long voyage. He is said to have lost many of the ships in the fight in the Strait, but still got his forces across – it sounds as though the warships were sacrificed to allow the cargo ships to cross with the army.[33]

When Pyrrhos finally sailed home to Epeiros in 274 he left Greek Sicily as divided as when he arrived, and Tarentum vulnerable to Roman conquest – which happened in 272. This brought all peninsular Italy into the Roman political system, and left Carthage dominant in Sicily. He is reported to have commented that he left a fine field of conflict for these two Great Powers. In maritime terms Carthage may have been dominant, but Syracuse still had a numerous fleet. There was no reason for Rome and Carthage to fight. The threat of Carthaginian conquest of Syracuse remained, but so long as Syracuse had its fleet the risk was small.

Chapter 8

A War for Sicily

In 264 Rome and Carthage fell into a war whose prize was Sicily, though it is usually called the 'First Punic War', reflecting the pro-Roman bias of both ancient and modern historians, the survival of Roman sources, and the disappearance of Sicilian and Carthaginian. The war lasted twenty-three years, the longest of any war in the ancient world, which indicates the balance of power between the two contenders. Given that the war was fought mainly in Sicily, it is fitting that the origin of the war lay in a dispute over a city in Sicily allied to Carthage but ruled by former Roman subjects.

The causes of the war and the course of its outbreak have to be outlined – it has scarcely been necessary for other wars mentioned so far – because the war was so unlikely. The two principals, Rome and Carthage, had not quarrelled seriously over any issue until this war began, indeed they had been allied in the fighting against Pyrrhos. The war resulted from the collapse of Agathokles' kingdom, which is why I have chosen a title for this chapter which emphasises this. It was also a war which was fought as much at sea as by land, so here we have, for the first time in this study, a case in which command of the sea was required for victory. It is worth remembering that this war took place at the same time that the Ptolemaic sea-supremacy in the east was at its height. Ptolemy II and III, the kings in office during the war, kept well clear, professing friendship for both states, and stern neutrality. Had they intervened, their sea power would have been decisive.

The war for Sicily began in a dispute over the city of Messana in northeast Sicily. It had come to be ruled by the Mamertines, a group of Campanian mercenaries, during the confusion after Agathokles' death. True to their origins they made a habit of raiding throughout the island, but eventually they found that Syracuse, which came under the rule of a new tyrant, Hieron, by about 270, was strong enough to defeat them and to confine them to their own territory. They thereupon made an alliance with Carthage, which had also gained control of the Lipari Islands since Agathokles' death.

Meanwhile, Rome took advantage of the death of Agathokles to expand its control over Magna Graecia. Tarentum was taken in 272, once the last of

Pyrrhos' garrison was withdrawn. Bruttium, the toe of Italy, was taken about the same time, though Rhegion proved to be troublesome. It was garrisoned by Roman troops, who, possibly influenced by their Italian contemporaries across the Strait, set themselves up as an independent city. An official Roman army came to recover the city, but found it difficult to do so without command of the sea; Hieron, who was already fighting the Mamertines, supplied the need, using some of Agathokles' old ships. Rhegion fell to the Romans in 269, the same year that Hieron beat the Mamertines. After his victory Hieron made himself king.[1]

Command of the Strait separating Messana and Rhegion, between the Ionian and Tyrrhenian Seas, was the vital strategic point, and this was no doubt in the minds of all involved. The result of the developments in 269 was that, in effect, Rome and Carthage faced each other across the narrow strip of water, each through a subordinate city ally. The wild cards in the area were, of course, the minor powers, the Mamertines and Hieron – Rhegion was under firm Roman control. Hieron's policy was presumably to recover the power and authority which the previous kings and tyrants in Syracuse had exercised in Sicily, and more immediately to suppress the troublesome Mamertines. These aims also imply a latent hostility to Carthage. He attacked the Mamertines again in 265 – no doubt they gave plenty of occasions for it – and when they were beaten, they appealed to their ally Carthage for help, and later they also sent an appeal to their Roman neighbours who controlled their original homeland in Campania.[2]

Both of the Great Powers responded, and in the process a clash between them developed. Carthage put a garrison into Messana, partly by promising Hieron, who was already at war with the Mamertines, that they would persuade the Mamertines to surrender.[3] Hieron withdrew when the Carthaginian garrison was installed, no doubt satisfied that these troops could control the city, though perhaps also frustrated at not gaining it himself – and being unwilling to indulge in a war with the African city.

The Romans sent one of the consuls for the year, Ap. Claudius Caudex, with instructions to assist the Mamertines, and with permission to make war on Hieron or the Carthaginians, or both. He clearly had virtual carte blanche, since the situation was obviously both complicated and changing. He marched to Rhegion, just across the Strait from Messana, with his consular army, which would be about 20,000 strong. He also gathered a fleet, an obvious requisite. The ships came from the city's 'naval allies', of whom Neapolis, Elea, Lokroi, and Tarentum are named, but there may have been others. The ships were mainly penteconters and triremes.

What happened next is described in several different ways by various ancient authors, none of them contemporary with events, and all of them subject to bias one way or another. There was fighting both at sea and on land, as a result of which Caudex gained control of Messana and installed his army there; both

Hieron and the Carthaginians withdrew their forces. It was his fleet, plus whatever cargo vessels he could find locally, which enabled Caudex to cross the Strait. He was opposed by a Carthaginian fleet stationed at Cape Pelorias, to the northwest. Some of the Roman ships were lost in the crossing, and at least one Carthaginian vessel, said to be a quinquereme, ran aground and was captured.[4]

The size of the rival fleets is never stated (and Hieron, active earlier, clearly stayed out of it). The Romans' own squadrons under the *duoviri navales* seem to have faded away, and are not heard of again after 282. Instead in 267 there were instituted four *quaestores classici*, whose task it was to see to the availability of ships to be provided by the coastal cities which were the naval allies. These officials were stationed at several places, of which Ostia, Cales in Campania, and Ariminum on the Adriatic are known; the fourth is not known, but one would expect him to be in the south, supervising Magna Graecia.[5] The total of ships called up was probably only ever small – in later Roman history the largest number ever produced by these allies at any one time was 24, in 191. It seems likely that this was the sort of fleet Caudex had at his disposal in 264. As a fleet, of course, it was unlikely to be very effective. There is no indication that the ships had ever operated or manoeuvred together, and none was larger than a trireme.

The strength of the Carthaginian squadron is similarly unknown, but it was probably small. Until 264 there had been no need for many Carthaginian ships in Sicilian waters, though it is probable that some were always on call at Lilybaion. Their squadron was brought to Cape Pelorias, and may have had a reserve base in the Lipari Islands, which the Carthaginians used in this way four years later. In peacetime the ships were usually taken out of the water to make them last longer. The main base for such storage was Carthage, where there were about 200 shipsheds in the naval harbour. There were similar facilities at Lilybaion, and perhaps at some of the coastal Greek cities in the Carthaginian area, but the capacity was probably not large enough to house a large fleet permanently in the Sicilian area – there was no need for that, anyway.

So the two forces at the Strait in 264 were fairly small – squadrons, that is, not large fleets. If the Roman fleet was twenty-four ships or less, the Carthaginian may have been even smaller. The basic difference between them, however, was not numbers, but size and weight. The Roman ships were undecked penteconters and triremes; the Carthaginian squadron, judging by the ship which ran aground and was captured, included quinqueremes. No wonder the Romans found it difficult to get across the Strait, which is only a couple of kilometres wide, though they did in the end succeed.

This Carthaginian quinquereme is the first vessel of this size to be mentioned in western waters since Dionysios I's time, a century before. In the wars of Agathokles nothing bigger than a trireme was used, though Agathokles' new navy, which he built up after 306, included quadriremes, at least one six, one nine, and possibly a seven. He is said to have been planning a new invasion of Africa when he died, so the new fleet was in part designed for that purpose. Pyrrhos' fleet of 200 ships included these vessels. He lost one of them at least, a seven, which is later found to be in Carthaginian hands; it may have actually been his own, brought from Epeiros. Agathokles' naval activity, and perhaps his long-term plans, would be no secret, and we may assume that Carthage was also making preparations to keep pace with his naval construction during the years of peace. Hence the quinquereme, a class which might well be able to cope with Agathokles' quadriremes. Agathokles' years of peace had seen, that is, an arms race in western waters at the same time as Demetrios in the east was busy outbuilding everyone else, and when Ptolemy II was building his great navy. In 264 the Roman trireme-and-penteconter fleet was therefore faced by a Carthaginian squadron which had been built to cope with Agathokles' ambitions.

In Syracuse, Hieron had the remains of Agathokles' fleet. If no new building had been done since his death in 289 the ships were by then reaching the end of their life, but Hieron had been able to provide ships to assist the Romans at Rhegion in 270, so it is likely that he still had a competent naval force in 264. This probably consisted, at least in part, of ships bigger than triremes, and this is presumably one of the reasons Rome was pleased to accept Hieron's alliance next year. Hieron's original alliance with Carthage had not brought him any rewards, only a Roman war. When the next Roman army landed at Messana in 263, it was commanded by both consuls, and so was perhaps 40,000 strong. At this, most of the Greek cities outside Syracuse joined the Romans. Faced by a possible siege of Syracuse, and the loss of his kingdom, Hieron quickly switched sides, despite the near presence of a Carthaginian squadron offering help.

The Roman naval force by the end of 263, therefore, was adequate for controlling the Strait, which was the absolute basic strategic necessity for conducting a war in Sicily. In 262, protected by the Syracusan alliance on their flank, with a secure base at Messana, and with the support, or acquiescence, of other cities in Sicily, the consuls were able to campaign right across the island, and to besiege and take Akragas. Akragas, independent until then, had become a Carthaginian ally in 264, when it looked as though it would not be difficult to throw the Romans out, and when all Sicily – Hieron, Akragas, Carthage – was united in detestation of the Mamertines.

The attack on Akragas had been countered by a Carthaginian army, which had been recruited in Sardinia, Africa, and Spain, and transported to Lilybaion. This force garrisoned the city, and more forces were brought from Africa during the siege, some being used to launch an attack on the besieging forces.[6] The Romans won the subsequent battle, but the Carthaginian garrison in the city escaped. The siege had lasted seven months, and was followed in the campaigning season of 261 by minor indeterminate operations which left both sides frustrated. The Carthaginian fleet was out, which may have reduced the number of forces available on land, though as a result many of the coastal communities in Sicily, which were vulnerable to seaborne attack, held to the Carthaginian side. The inland communities, which could be reached by the Roman army, chose to remain with the Romans. It does not seem as though anyone in Sicily was enthusiastic for either of the outsiders.

The Carthaginian fleet was also used aggressively, mounting a number of raids on Italy.[7] Which places exactly were attacked is not stated, but the Greek towns along the northern Calabrian coast and along the 'shin' of Italy would be the obvious targets. Apart from the fact that it was from places such as these that Rome recruited its warships, it might have been thought that they would follow the example of the coastal cities of Sicily and defect to the Carthaginian side. Their inclusion in the Roman commonwealth was scarcely a decade old.

The one result of these raids was that the Senate in Rome decided to develop a Roman fleet. This was not a sudden decision – there had been a small Roman fleet over fifty years before, and the naval allies provided ships to operate under Roman command. It was, however, a major change in scale. The conqueror of Akragas, M'. Valerius Maximus, is said to have remarked, after the victory against the Carthaginian relieving force, that to conquer the island it was necessary to have a fleet.[8] This was scarcely an original insight, since even the slightest acquaintance with Sicilian history reminded anyone that the Greeks had always finally failed in attempts to capture Lilybaion, which was open to relief by sea. By surviving repeated sieges, the city repeatedly became the base for Carthaginian conquest or reconquest. Pyrrhos' failure to take the place only fifteen years before was only the latest example. But Valerius' comment, if accurately reported, does mean that the Romans were seriously thinking of the total conquest of the island, even at this early stage of the war – for he was scarcely the only Roman with such ideas, and the Senate did agree to develop the fleet. Again, acquaintance with Sicilian history would suggest that only by removing the Carthaginians totally from the island would Rome enjoy undisturbed possession.

The object of the war, on the Roman side, had thus clearly very quickly developed from merely rescuing the Mamertines to a determination to conquer the whole island. This may be mere ambition among the annual consuls, but it may also have been fuelled by a sense of obligation: many Sicilian communities had joined the Romans voluntarily, and if the Romans now merely made peace, while both Hieron and the Carthaginians kept a presence on the island, those communities were liable to suffer.

The Senate's decision was to build a hundred quinqueremes and twenty triremes.[9] The story is that this was done in sixty days, and that the rowers were trained on land while the ships were being built. It is also claimed that no one knew how to build quinqueremes, so the captured Carthaginian ship was copied.[10] How much of this story is true is difficult to decide, just as it is hard to understand where the ships were built, and how they were crewed.

The number of ships was presumably decided in the knowledge of the numbers of the existing Carthaginian forces, of which a fleet of fifty ships is mentioned next year.[11] This was not the whole Carthaginian fleet, but probably was the larger part, so 120 Roman ships was clearly more than adequate to meet the present enemy – reinforcements could be expected also from Syracuse, and from the naval allies. Training rowers on land is quite feasible and is attested elsewhere. It does imply that the men recruited for the service were landsmen: a mutiny a year later by a group of Samnites from the Apennine Mountains who had been conscripted for sea service is an example of the sort of men who were taken.[12] It does not seem likely that it was too difficult to train men as rowers, so using landsmen and training them on land is all quite reasonable.

The real difficulty with the story is the implication that it was all done so quickly. Speed was undoubtedly of the essence, and it was obviously necessary to have the full fleet available from the start of its cruise, otherwise it could be destroyed in penny packets. Yet one must assume a certain lapse of time between the decision and its implementation, and a further space of time before completion. There was also probably a period of time between the decision to build and the decision on *what* to build. So the implication that only sixty days passed in all this is highly unlikely.

The decision to build a fleet was made soon after the news arrived of the conquest of Akragas. That city fell early in 261, perhaps in January, and so the Senate's decision was made at the latest in February that year. In effect that gave a year for the work to be done, since the fleet did not put to sea until some time in 260. The selection and cutting of timber could begin right away, and the decision as to the size of ship could be put off for a time while arrangements were made as to where the ships would be built and who by. Several places were probably the building sites, for building so many ships in one place seems unlikely and would certainly be uneconomic. The builders would need to be skilled shipwrights, who were not numerous, but such men did live in every port in Italy. It would clearly be best to take the wood and the work to the men, rather than concentrate everything in one place, which would soon become overcrowded and inadequate.

The choice of the quinquereme as the basic type was probably not difficult, since that was apparently now the standard Carthaginian type, as it was in the navies of the eastern Mediterranean. Building bigger ships would take longer, but building smaller ones would only produce ships vulnerable to the Carthaginian quinqueremes. It may well be that no quinqueremes had ever been built in Italy before 260, though there were bigger ships at Syracuse, and perhaps at Tarentum. But the quinquereme is only an enlarged trireme, a type of ship which was common in all the ports of Italy. No doubt the captured Carthaginian ship was carefully examined and its lines could be copied, but that Italian shipwrights were unable to build such a ship without copying another must be doubted.[13]

The ships were manned as they were launched, and the rowers will have had further seagoing experience before the order was given to sail to Messana; seventeen of the ships were taken there in advance by the new consul Cn. Cornelius Scipio.[14] So the rowers had quite a quantity of experience on land and at sea by the time they reached Sicilian waters. What they did not have was any serious experience in fighting at sea. This became clear when Scipio took his seventeen ships to Lipara hoping to seize that place by treachery or surprise. News of his move reached the Carthaginian commander Hannibal at Panormos. He sent a Carthaginian senator, Boodes, with twenty ships to take care of it. When Boodes unexpectedly attacked the Roman forces at Lipara at dawn the crews fled ashore; Scipio was captured.[15]

This was perhaps not a fair test, since the Roman ships were trapped in the harbour and could not be manoeuvred; and an attack at dawn on inexperienced troops was almost guaranteed to cause panic. (Certainly Scipio was not blamed; he was re-elected consul six years later, despite being awarded the soubriquet 'Asina' – 'she-ass'.) But Hannibal's confidence grew. He took a fleet of fifty ships to meet the new Roman fleet, with the likely intention of disputing its progress and perhaps destroying it completely. Neither knew of the other's presence, but Hannibal was the one who recoiled when they met. This is not surprising, for he was outnumbered by two to one, which was probably unexpected, and the Romans were in a good formation. Hannibal escaped, having lost some ships. The Roman fleet had, even as early as this, thereby secured command of the sea of northern Sicily, and of the Strait, at least.[16]

The other consul, C. Duilius, took command of the fleet, which was now, if not earlier, equipped with the *corvus*, 'raven', a hinged gangplank designed to be loosed to fall on the enemy ship and so pin two ships together, allowing Roman soldiers to board the enemy.[17] The Carthaginians brought up reinforcements and stationed their fleet, now 130 strong, at Cape Mylai, twenty kilometres west of Cape Pelorias. No doubt the Carthaginian fleet included the seventeen ships captured from the Romans. Seeing that the *corvus* in the coming battle was a surprise, the ships captured at Lipara cannot have had this device, which must therefore have been fitted in Sicily as a last-minute addition. One wonders whose idea it was, but it must have been Duilius who approved the idea and ordered its installation.

The size of the Roman fleet which sailed westwards from Messana is not stated. Of the 120 ships originally built, seventeen had been lost at Lipara, but some of the Carthaginian ships had been taken in the earlier encounter, and there were ships from Rome's allies already at the Strait. Probably, therefore, the

Romans had more or less equality in numbers of ships with the Carthaginians, 130 each, but fewer of them were quinqueremes. It was the *corvus* which made the difference.

The Carthaginians, despite their defeat earlier, were overjoyed to find the Roman fleet advancing. Overconfidence as usual proved costly. The Carthaginians advanced in some disarray, charging right in without manoeuvring. The *corvus* came down when they got to close quarters, and the thirty foremost Carthaginian ships were quickly captured, including a seven which was King Pyrrhos' old ship being used as the flagship. The rest now manoeuvred to attack more carefully but lost another twenty ships before the remainder (eighty ships) retreated; it was perhaps in this phase that the Carthaginians lost some ships sunk, which was not something the *corvus* could achieve. So we must envisage this second phase as a much more traditional fight: that is, the Carthaginians, by using normal manoeuvring and ramming tactics, could neutralize the *corvus*. The Romans finally realized this, and the device was abandoned.[18]

The victory had few immediate results. The Carthaginian fleet had not been destroyed, and could be reinforced up to its original strength. The Roman intention was still to conquer the island; Duilius switched to the land campaign after his naval victory. In the longer term, however, the Roman victory allowed them to consider far wider strategic possibilities. Next year, one consul campaigned in Sicily, resisting a strong Carthaginian counter-attack, while the other operated in Corsica and Sardinia. This latter campaign will have required a substantial proportion of the Roman fleet, which could therefore not be used in Sicilian waters, except to control the Strait. The consul L. Cornelius Scipio (Asina's brother) first eliminated the small Carthaginian presence in Corsica, then sailed for Sardinia. He met and frightened off a Carthaginian fleet, which retreated, presumably because it was outnumbered. When he reached Olbia, in northeast Sardinia, however, he met a reinforced Carthaginian fleet, and so he himself turned back.[19]

The Carthaginian commander in Sardinia was the man, Hannibal son of Gisgo, who had commanded at Mylai (and who had earlier commanded at the defence of Akragas). This probably means that he was able to use the Carthaginian forces in Sicily to resist the Roman advance in Corsica and Sardinia, operating on what in land warfare is usually described as interior lines. The major Carthaginian bases – Carthage, Lilybaion, and Sulci in Sardinia – were all within relatively easy reach of each other. The Romans, however, operating from Rome (or Ostia) and Messana, were condemned to long voyages to reach the scene of warfare. The campaign in Corsica and Sardinia was probably therefore aimed at taking over one of the Carthaginian bases, Sulci, and so gaining a base closer to Lilybaion and Carthage.

The next year, 258, the consul C. Sulpicius Paterculus repeated and extended Scipio's campaign. He had a substantial land force, perhaps his full consular army, and he captured a number of Carthaginian posts in Sardinia, which were mainly on the coast. Hannibal son of Gisgo was eventually induced to bring out his fleet when the final post, Sulci in the south west, came under threat. From this one may conclude that he had not done so earlier because he was outnumbered, probably because Sulpicius had brought a larger fleet with him than the year before, or because some of the Carthaginian ships had been sent away to Sicily or Carthage. The subsequent battle, fought in a fog, resulted in Hannibal's defeat; he got ashore, but was killed by his own men. Sulci presumably fell to Sulpicius, thus gaining Rome a considerable strategic advantage, though Carthaginian mercenaries continued to operate in the island. This all implies that Rome was interested mainly in the naval base, not in the island itself. The naval victory also implies that the majority of the Roman fleet strength, apart from the naval guard kept at the Strait, was devoted to this two–year campaign.[20]

This was, in effect, the Roman naval exploitation of the victory at Mylai. The operations in Sicily in those years were inconclusive, despite a Carthaginian victory at Thermai in 259. The Roman strategy would seem to have been to hold the Carthaginians in play in Sicily, while employing their naval power elsewhere. But Sicily was the main battleground, the explicit prize in the war, and peripheral operations such as those in Corsica and Sardinia, would not have much effect – unless they became so threatening that the Carthaginians were distracted. In Sicily, however, victory would be won only by the conquest of every town and city and fortress by Roman arms – a necessity which these peripheral operations were no doubt intended to evade. For Carthage, which was employing its well-tried strategy, learned from frequent past Sicilian wars, victory would be achieved just by holding on to one fortress. Rome had to be aggressive; Carthage could win by a stubborn defence.

In 257 both consuls operated in Sicily, implying that the Sardinian operation had been closed down, having succeeded, but having been seen to lead nowhere. The Roman fleet returned to Messana. One of the Carthaginian naval bases was at Lipara, which had been attacked by a small Roman force the year before, without success. Lipara's geographical position made it difficult to push Roman forces along the north coast of Sicily, since a fleet based at Lipara could interrupt a Roman attack – as would be seen in this campaign. Now that the full Roman naval power was available, Lipara was again on the agenda. The Carthaginian fleet there was about eighty ships; the Roman fleet at Messana is said to have been 200 ships. This may be an exaggeration, but if no ships had been lost since Mylai, and some had been taken at Sulci it is possible that, putting quinqueremes and triremes and allied ships and the captured ships together, a total approaching 200 is possible. Many of them, however, were lighter ships which could not be employed in the line of battle.

The Roman fleet sailed west along the north Sicilian coast, and was intercepted off Tyndaris, about fifty kilometres from Cape Pelorias, by the Carthaginian fleet. Several versions of what happened have survived: either the Carthaginians caught the Romans divided, or a small Roman force was sent in advance as a lure, or the consul C. Atilius Regulus attacked impetuously, but an initial Carthaginian success against a small Roman squadron was reversed when the full Roman fleet came up. It does seem unlikely that the Carthaginians would attack a fleet more than double its numbers, so a divided Roman fleet would have seemed a good target. The losses on both sides were relatively minor – nine Roman ships sunk, eight Carthaginians sunk and ten captured. Strategically it was a defensive victory for the Carthaginians, since the Roman aim had been to attack Lipara, and this was abandoned. Instead Atilius took part of the fleet to raid Malta. But the Carthaginian losses had been a quarter of their fleet, while those of the Romans only five per cent.[21]

The size of the Roman fleet at Tyndaris presupposed that soon it would gain complete control of the seas around Sicily. Another fight like Tyndaris would reduce the local Carthaginian fleet to a negligible size, if any of their ships survived at all. So, one result of this battle was to stimulate Carthage into building up its naval strength again, with the aim of gaining full naval mastery of the Sicilian seas. At the same time the raid on Malta may have stimulated thoughts at Rome of another peripheral action, an invasion of Africa. This possibility had been rumoured during the Sardinian campaign, and no one on either side can have been ignorant of the exploits of Agathokles in Africa fifty years before. It may have seemed a war-winning notion, which would compel Carthage to bring its Sicilian forces home, leaving their positions in the island vulnerable, or even bringing peace without having to fight in the Sicilian trenches. (If so, the lesson of Agathokles' failure had been eclipsed by the recollection of his audacity.) But a fleet of a hundred-plus ships was not enough for such an enterprise, so, on the Roman side, a further result of Tyndaris was to stimulate more shipbuilding, just as it was, for different reasons, at Carthage. The fleets which met in battle next year, then, were made up partly of experienced ships and crews and partly of new ones on both sides. The numbers were huge: 330 Roman and 350 Carthaginian ships – though there is a dispute about both of these numbers, largely based on incredulity.[22] Even more impressive are the numbers of men involved: 140,000 Romans and perhaps 150,000 Carthaginians, but this is scarcely noted. The Roman ships certainly carried a large complement of soldiers as well as their crews, for there seems to have been no cargo or merchant vessels with the fleet; Polybios notes that four legions of soldiers were embarked, 39,600 men. Both sides were putting out their maximum effort.

The Roman fleet sailed from Messana and round the southern cape of Sicily, then along the south coast as far as Cape Eknomos, where the soldiers embarked, having marched across the island. The Romans' intention was to transfer that army to Africa, and then most of the fleet would return to Sicily. The embarkation, at a place which was an open beach and roadstead, took some time. This allowed time for the Carthaginian fleet to come up. The intentions of the Carthaginians are rarely examined, but in a fleet with 150,000 men on board, it is evident that 50,000 or so were soldiers. How obvious the Roman intentions were to the Carthaginians is not clear, though one would imagine that it was fairly well-known amongst the Romans in Sicily that they intended to invade Africa, and so it is very likely that the word had reached the Carthaginians in the island. So the Carthaginian fleet had sailed along the south coast of Sicily looking for the Romans, though that was probably not the Carthaginians' original plan for this campaigning season. Since they had so many soldiers on board they were probably intending a landing somewhere in Sicily, or possibly even in Italy. An attack on Syracuse would be in keeping with their naval tradition. At the same time, they now knew that Roman naval tactics were centred on boarding and that their ships carried large numbers of soldiers for that purpose; to combat this the Carthaginians also needed large numbers of soldiers on their own ships. So they, like the Romans, were prepared both for a naval battle and for a landing. The former happened.

The naval encounter took place as the Romans began their voyage towards Africa, but it was not unexpected. Both sides had their fleets in formations which had clearly been thought out with a battle in mind. The Romans had four squadrons (they called them 'legions'), of about equal size: two, each led by a six with one of the consuls on board, sailed in line ahead *en echelon*, forming lines which gradually spread out to either side from the leaders, each ship acting as protection for the one in front; a third squadron formed the third line of the triangle, and some of the ships here were towing horse transports; a fourth squadron was behind this line. The Carthaginians had three quarters of their ships in a long line, sailing at right angles to the coast; the rest were in line ahead close to the coast; their formation thus formed an L-shape.

Seeing the enemy, the two forward Roman squadrons were led directly ahead to drive at the centre of the main Carthaginian line. Close to the coast, the fourth Carthaginian squadron ignored this fight and went on to attack the Roman third line from the coastal side. The main Carthaginian line was thinner in the centre than on the wings, and withdrew before the Roman charge. The weight of the Roman attack lay in the centre, with the two sixes (the only time the Romans in this war used a ship bigger than a quinquereme), and by penetrating the Carthaginian line at the head of their expanding lines, the plan was that the main enemy force would be broken into two parts. The great triangle's head drove into the line, which retreated and bent, so that the triangle was menaced on both sides by the Carthaginian wings. Each of the two forward Roman 'legions' had about eighty ships; they were about to be attacked by two larger squadrons of perhaps 120 ships each. But this is what the Carthaginians expected. As the Carthaginian line was bent back, Hamilcar the commander gave the order to turn and attack. At the same time a part of the Carthaginian right was detached. It imitated the action of the fourth squadron on the left and raced ahead to attack the rear Roman line from the seaward flank. All this had clearly been worked out in advance, and made use of the more skilful Carthaginian oarsmanship, for they were always faster and more manoeuvrable than the Romans in these battles.

The battle thus resolved into three separate fights, the two flank attacks on the forward Roman 'legions' and the main one at the head of the Roman formation. These were fights of manoeuvre, which was the Carthaginian skill; but the Roman manoeuvring turned out to be just as skilful. The fight at the head of the Roman formation was the first to be resolved, for Hamilcar found himself defeated and fled the field. This was the decisive moment. It could be that Hamilcar's squadron had been weakened by the detachment of the flying column sent to attack the Roman fourth line, by which the Roman squadron became the stronger, but it was a major achievement to beat such a Carthaginian force in open battle. One of the consuls, L. Manlius Vulso, took charge of the battlefield, collecting captures, rescuing sailors, and no doubt waiting to see if this Carthaginian retreat was another manoeuvre rather than a defeat; the other consul, M. Atilius Regulus, turned back to assist the third and fourth lines of the fleet in their fights. The third line in particular had been penned into a narrow area close to the beach, though this did allow the use of the *corvus*, which had been impossible in the other fighting. Atilius' arrival now surrounded the Carthaginian surrounders, who mainly surrendered.

The losses on both sides are instructive. The Romans lost 24 ships sunk, the Carthaginians thirty, no doubt mainly in the fight at the head of the Roman column. The Carthaginians lost another 64 ships captured, and these latter were mainly from the squadron forced to surrender at the end of the battle, which had had at least fifty ships. So fourteen were captured in the rest of the fighting, presumably mainly by the use of the *corvus*, while the two sides were about equal in sunk vessels, which was normally caused by ramming. It does not look as though the *corvus* was a particularly effective battle winner in this fight. The threat of its use had kept the surrounded Roman ships of the third line from being beaten, but they were certainly losing until Atilius arrived.[23]

The Roman fleet put back to Messana to repair and recover. The Carthaginians retreated along the coast to their base at Herakleia Minoa. The former were now about 300 strong, to which the 64 captured ships could be added if they could be manned, but considerable repairs were needed to the damaged ships.[24] The Carthaginian fleet was still over 250 strong, though again some considerable repairs had to be done. The Roman victory was therefore by no means decisive. It did, however, permit the Romans to make a second and successful attempt to sail across to Africa, though only after some delay. The Carthaginians at Herakleia Minoa were in a good position to intercept the Romans if and when they tried again. From the Carthaginian viewpoint, there was no guarantee of another Roman attempt, since more than once the Romans had fought a battle, won (at least by the Roman lights), and then retired – Tyndaris the year before was the latest case. Hamilcar was recalled to Carthage, and most of the ships with him. The odds were too much against them, and so the Carthaginian government had evidently decided to meet the coming attack in Africa.

So when the Romans set off again they had a clear passage. An advance force of thirty ships went first, to scout out the target, and the rest followed. The Carthaginians feared that the invasion would be targeted directly at the city, and they kept their fleet manned and ready to meet that attack. But the Romans went to the east of Cape Bon and landed at a place called Aspis, where they entrenched a camp, and then sent home for instructions.[25]

This seems an extraordinary initial reaction by the Romans to their successful landing, but it is perhaps a consequence of the Eknomos battle, which clearly imposed a considerable delay on the Romans. This invasion had been subject to repeated delays – in embarking the troops at Eknomos, in the battle, in the repairs at Messana after the battle, and in the voyaging to and fro which will have consumed several days each way, and now for the search for instructions. (The messenger had to go to Rome, discussion had to take place and a decision be reached, then the messenger had to return to Aspis.) New instructions were needed because the season was now so late that communications between Sicily and Africa would soon be cut by winter.

The result bears all the signs of a compromise: one consul, Atilius, should remain with 15,000 troops and forty ships in Africa over the winter, and the rest should be withdrawn to Sicily.[26] It was a recipe for disaster, and this duly happened. The main fleet evacuated the surplus men to Sicily, and Atilius began to campaign. The Carthaginians were defeated at first, but revived their army by hiring a Spartan, Xanthippos, to drill and command it, and he brought the Roman army to defeat in the spring. In Rome a new expedition was in preparation, but this was used to evacuate the last of the Romans who were holding out in the Aspis camp. A fleet of 350 ships were sent; it was met at Cape Bon by a Carthaginian fleet of 200, which was defeated with considerable loss. The numbers in this episode are very confusing, but it seems that Carthage's fleet was substantially reduced as a result, perhaps to only a hundred ships or less (apart from some other ships which had been kept in Sicily all the time).[27] The Romans landed, but found that, as a result of the ravaging committed by the Romans, there were not enough supplies available to feed their men (over 150,000 soldiers and sailors) and so the whole lot were taken on board the ships to return to Sicily. Making landfall at Kamarina, the fleet was struck by a storm; only eighty of the ships survived; perhaps 100,000 men died.[28]

This disastrous result – defeat, evacuation, destruction, mass death – was very largely a result of the Eknomos fight. The long delay imposed by having to prepare for, and recover from, that battle was the direct cause of the Roman decision to keep the main fleet in Sicily over the winter. There were still 250 Carthaginian ships available, at least, and it will have become clear that the clash at Eknomos was actually between two invasion fleets. Anyone who knew anything of Agathokles' history knew of that weird moment when each side was besieging the other's city. If the Roman army was busy in Africa, there was nothing in Sicily to stop a major Carthaginian attack by land, perhaps even on Syracuse, or still worse, from the Roman point of view, on Messana. The Roman forces were not strong enough to hold Sicily and invade Africa at the same time. Thus it followed that Atilius' army in Africa was beaten, and the Roman fleet was destroyed, despite the commanders being warned, explicitly and repeatedly, of the danger of a storm at that season.

The survival of only eighty Roman ships left the Carthaginians with the stronger fleet; the deaths of 100,000 Roman soldiers and sailors left the Carthaginians stronger on land as well. But the Carthaginian strategy remained wholly defensive. Both sides set about rebuilding their fleets. Carthage spent the next year recovering control in its African hinterland, for the Roman campaign had stimulated local revolts; this led on to a larger series of land conquests, which took ten years or so, and produced a change in Carthaginian attitudes to the Sicilian war. Briefly, Africa had now become more important than Sicily. In Sicily a commander called Carthalo retook Akragas, but could not hold it, partly because he had little or no support.

At Rome the failure to win the war by peripheral campaigns in Sardinia and Africa brought full attention back to Sicily. 220 new ships were built in 255/254 to bring the Sicilian fleet up to 300 ships; Scipio Asina was re-elected consul for 254.[29] He used the fleet in combination with the army, and made, for the first time in the war, a serious advance along the north coast. Previous attempts had used the fleet alone, and had been disrupted by the Carthaginian naval force stationed at Lipara. Scipio Asina, assisted by his colleague, another re-elected consul, A. Atilius Caiatinus, ignored that Carthaginian fleet and campaigned right along the coast. Part of the force, under Atilius, failed to capture Drepana at the western end, but Scipio's fleet and army took Panormos and Kephaloidion. As a result several other Carthaginian-dominated towns came over to the Roman side.[30]

For the first time since the capture of Akragas at the beginning of the war, an imaginative strategy had been implemented, and taking Panormos involved a major advance in Roman control of the north coast. It was clearly Scipio's strategy. He stayed on next year as proconsul, but one of the consuls for the year (253: C. Sempronius Blaesus) was of the African persuasion and took the fleet on a long raid to the African coast. He got into trouble in the Gulf of Syrtis where the unusual local tide almost caught him. The fleet ran back to Sicily, sailing round the western end of the island, flaunting its presence and power before the Carthaginian towns there, called in at Panormos, and then sailed back to Italy, apparently heading for Rome (or Ostia). Off Cape Palinurus it was caught in a storm; 150 ships were wrecked. As usual, the consul survived; thousands of his men died.[31]

This sequence of successes on land and failures at sea finally convinced the Senate to change course. The lost ships were not replaced, and the war in Sicily reverted to a campaign against individual cities, which was the only way the Romans were ever going to win. In 252 Thermai, another of the northern coastal cities, was taken, and, at last, Lipara. Carthage now had to send more forces to Sicily, and in 250 made an attempt to retake Panormos, which failed. Carthage's version of the *corvus* as a surprise weapon, their elephants, was unsuccessful here. Rome's control of all eastern and central Sicily had now driven the Carthaginians back to their western redoubt, the cities of Drepana and Lilybaion in particular, both of which were significant naval bases. Other cities were evacuated – Herakleia Minoa, Eryx, Selinous – and their populations were concentrated into the two bases. In 250, after increasing their fleet by fifty new ships, the Romans finally grasped this awkward nettle.[32]

The two consular armies laid siege to Lilybaion, assisted by a fleet of 200 ships. Most of the ships were beached, and so the Carthaginians were able to run in a fleet of fifty ships with reinforcements and supplies, and bring them out again without any interference;[33] later a particularly skilful captain, Hannibal 'the Rhodian', using a light and swift ship, probably a quadrireme, repeatedly ran the Roman blockade, until challenged and defeated by another quadrireme manned by a picked crew.[34] The siege was therefore slow, lengthy, and unsuccessful. The main Roman difficulty – apart from the resistance of the besieged – was in keeping their own forces supplied, so they had to reduce the army in the siege lines to a single consular army, which was not large enough to force a decision. Instead of direct attacks on the city, the siege was replaced by a blockade by land. The Carthaginian control of Drepana and of the Aegates Islands enabled them to send in supplies, and, by seaborne raids and cavalry raids, to interfere with the Roman supply lines. In 249 the Roman fleet made an attempt on Drepana, but the Carthaginian ships came out, forced the Roman fleet back against the shore and won a major victory, capturing 93 Roman ships; the consul escaped with just thirty ships.[35]

The Carthaginian commander in Drepana, Adherbal, had already used his ships to raid along the Sicilian coast; now that he had destroyed the majority of the Roman fleet in the west he would be free to raid further and more often. Soon after the battle a further seventy ships arrived from Carthage, commanded by Carthalo (who had earlier taken Akragas for a time). He almost at once went to Lilybaion to try to finish off the Roman naval forces there, and succeeded in capturing five ships and destroying others. Adherbal had started the battle at

Drepana with 130 ships, and had captured 93. Now he gave Carthalo thirty more ships, so dividing the Carthaginian force more or less equally, and took the captured ships and crews off to Carthage.[36]

Meanwhile another Roman fleet had been built. It was sailed to Messana and joined with the ships there. Under the command of the consul L. Iunius Pullus, a fleet of 800 transports with supplies for Lilybaion was assembled, and Pullus intended to escort them with his 120 warships. He sent some transports on ahead, escorted by some of his warships under the command of quaestors. He had not yet heard of the loss of the Roman fleet at Drepana, but he assumed it was too risky to sail past that city, so the two Roman supply convoys sailed south from Messana, round Cape Passaro and along the southern coast. The first convoy, only lightly guarded, met Carthalo and his hundred warships near the site of the Eknomos battle. The Roman quaestors, hopelessly outnumbered, quickly headed for the shore and beached their ships. They gathered up some artillery from the town of Phintias nearby, and prepared to defend their ships and the convoy with this artillery from the shore. Carthalo captured a few transports, but flinched at fighting the artillery. He retired a little way, waiting for the convoy to sail on. The second convoy, escorted by many more warships, approached. Carthalo could not let the two forces join, so he sailed past the first and stationed himself between them. There may have been a fight, though this is not certain. What is certain is that Carthalo was warned by local experts that a storm was coming. He sailed away, and so when the storm arrived it struck only the Roman forces. The result was the total destruction of the fleet and the transports – only two ships are said to have survived; the consul, as usual, also survived; casualties must have been, once more, in the tens of thousands.[37]

This left Carthage once more supreme at sea. The Romans once again concentrated on the land war. The blockade of Lilybaion continued, and Drepana was now besieged as well. Carthalo used his fleet to raid Italy, but the Romans made no attempt to rebuild their fleet.[38] Both sides settled for the status quo, the two towns remained under siege, and the Carthaginian fleet was free to roam and raid.

A new commander, Hamilcar Barca, was appointed to Carthaginian Sicily, apparently as a commander-in-chief over both naval and military matters. At Drepana the Romans captured an island off the harbour, but could make little use of it. Hamilcar was, by contrast, able to raid the Italian mainland in Bruttium and the city of Lokroi. The Romans licensed privateers to make raids on Africa. Hamilcar seized a defensible promontory, Heirkte, on the north coast as an advanced base from which to mount raids on the rest of the island, and by sea. A naval raid as far as the east coast of Sicily is recorded, and another along the Italian coast as far as Cumae in Campania – there were no doubt others.[39] After three years Hamilcar abandoned Heirkte for Eryx. All this was only minor warfare, and none of it could be expected to bring victory. The Carthaginians' continued warfare in Africa implies a loss of interest in Sicily, but the continued successes of Hamilcar would clearly prevent any peace except one which allowed them to hold on to Lilybaion and Drepana. For Rome the longer those sieges went on the more it was politically necessary that they be concluded successfully. Stalemate.

It may have finally dawned on the Romans that, despite Hamilcar Barca's activities, Carthage had largely lost interest in Sicily. The besieged cities had been kept supplied, the fleet and Hamilcar had conducted raids, but no real attempt had been made to recover control of large areas of the island. From 250 onwards, Carthage's investment in the Sicilian war was limited to Hamilcar Barca's exploits and occasional relief expeditions to Lilybaion and Drepana. Meanwhile Rome maintained two full consular armies at the sieges, and in doing so suffered substantial casualties, mainly from disease. In such circumstances Rome could not win. The options were, in the face of the Carthaginian defensive strategy, to capture the cities, defeat Hamilcar, or regain command of the Sicilian seas. This last option was paradoxically the easiest. Hamilcar appeared unbeatable, and even if he was defeated, he could very likely set up his raiding camp elsewhere. Seven years of siege brought neither of the cities within reach of capture. Only by cutting off their supplies could the garrisons be weakened to such an extent that they became vulnerable. And that meant gaining command of the local seas.

So a new Roman fleet was built. Evidently not everyone was convinced, for the ships were built by private subscription. There had also been some serious thinking being done, for it was decided to abandon quinqueremes, which the Romans had never been able to manoeuvre easily and which had always been slow. Instead the captured ship of Hannibal the Rhodian, which was probably a quadrireme, was taken as a model. Such ships were lighter than quinqueremes, more manoeuvrable, required smaller crews and, perhaps crucially for the subscribers, they were cheaper to build.[40]

Some 200 ships were constructed. The consul Q. Lutatius Catulus – who appears to have been the driving force behind the whole venture – took the fleet to Sicily in 242, reinvigorated the siege at Drepana, and spent much time training and exercising the crews and the ships. By surprise attacks he seized control of the harbour at Drepana and of the beaches at Lilybaion, both of which actions signalled an early tightening of both blockades.[41] Carthage replied by sending its main fleet to run in supplies, but did not do so for several months. No doubt it took time to realize the implications, and for the garrisons to run down their stockpiled supplies. There may well have been difficulties in building and manning the ships needed, and the government may also have been reluctant to invest in yet another great fleet. In the event 250 ships were sent, but the crews were inexperienced, and it was intended to take on board soldiers actually in Sicily. That is, the Carthaginian plan was to get into Lilybaion, unload the supplies, collect the soldiers, and then come out to fight.

Hanno, the commander, took his fleet to one of the Aegates Islands, Hiera, to the west of the besieged towns, and waited for a fair wind to sail into Lilybaion.[42] Lutatius took his fleet to another of the islands and lay in wait. When a westerly breeze blew, the Carthaginians set sail, using their sails. Despite the uncomfortable sea the Roman fleet came out to dispute its passage.

The whole situation was the reverse of earlier encounters – the Carthaginians outnumbered the Romans 250 to 200, their crews were inexperienced whereas the Romans were well-trained, and in the fight the Romans kept to the high seas and pushed the Carthaginians against the land. But above all the main reversal was that the Roman ships were more mobile, faster, and more manoeuvrable. The Carthaginians lost fifty ships sunk and seventy captured, which effectively halved their fleet. At least some Roman ships were damaged and some sunk, but as with most ancient battles the victors suffered far less than the losers.[43]

Hanno went back to Carthage with the news. The Carthaginian government, pausing only to crucify Hanno himself for his failure, at once sent plenipotentiary powers to Hamilcar Barca. He and Lutatius, both victorious, sparred for a time over some of the terms of the peace treaty, but eventually agreed that Lilybaion and Drepana and Eryx should be evacuated by the Carthaginians, and that Sicily should be at Roman disposal.[44]

The war had reached its twenty-fourth year, an appallingly long time. It had cost several hundred ships, and several hundred thousand lives. The census count of Roman citizens fell by 140,000 between 263 and 243, both of them war years, but most of their casualties were not citizens, for it was those poorer and less privileged than the citizens who were the cannon fodder in this war. So, if the number of citizens fell by 140,000, the number of the poorer conscripts probably fell by three times that, or more. As a casualty rate, it approaches Great War proportions.

It was at sea that these casualties occurred. Until 264 Roman military might had been confined to the land, and casualties in its wars were bad enough – consuls in command were prodigal with lives – but once the city launched its first fleet, the numbers of dead climbed. Many of the casualties were the result of ignorance and incompetence. The annually elected consuls only rarely showed any naval, or even military, prowess. Indeed, only Q. Lutatius Catulus, who brought the war to an end, showed any signs of a seaman's intelligence. It was only in the final Aegates Islands fight that we can see Roman ships acting as normal warships, though there had been some signs of it at Eknomos; even there, when the Romans called the several formations in that battle 'legions', it was a clear sign of their land-mindedness. In every other battle they tried to make sea battles into imitation land fights. The *corvus* was their device for this, and it worked only once, at the first battle at Mylai. It was a typical landsman's solution to a problem, and fits in well with the hit-and-miss Roman strategy which tried to win the war by peripheral operations in Sardinia and Africa rather than buckling down to the real work – the *corvus* was an attempt to avoid the hard work of learning sea warfare. By the end it had been discarded, probably because it made the ships unseaworthy, as well as being useless most of the time: the huge casualties in the storms were partly the result of the unbalancing effect of the *corvus*; the other major factor was the arrogant unwillingness of the consuls to hear any local weather advice.

The point is made by one of our main sources for this war, Polybios, that for length and damage, there was nothing in the wars in the eastern Mediterranean to compare with this war, and he contrasts this war explicitly with the campaigns of 'Antigonos, Ptolemy and Demetrios' in the east.[45] What he does not point out is that the length of the war was largely due to the general equality of power between the two sides. It was not that Rome and Carthage were generally equal in manpower and resources, still less in military ability. Rome had greater supplies of all these than Carthage. And Rome clearly knew this, going into the war blithely assuming that the city could, without too much effort, defeat the coalition of Carthage and Syracuse. In this they were quite right, except that the Senate decided to turn to the sea.

To win this war, such a development was quite unnecessary. The conquest of the several cities of western Sicily was all that was required. It was not even necessary to conquer them all: to control Sicily, with all the resources of Italy behind it, Rome did not need to possess either Lilybaion or Drepana. Once Rome began producing fleets, it dissipated its real power, which always lay in its army. For, given a fleet, the temptation always existed to launch expeditions to Corsica and Sardinia and Africa, none of which were at all relevant to the war in Sicily.

Rome's use of sea power was therefore expensive and, until the last battle, based on mistaken premises. Making encounters at sea into imitation land battles would not achieve command of the sea. To do that seamanship within the high command was required. The battles of the first part of the war – Mylai, Tyndaris, Eknomos – were indecisive for that very reason, for in each case most of the enemy vessels survived. The sensible use of sea power was shown by the Carthaginian ships based at Lipara, where, in the battle at Tyndaris a fleet of eighty ships fought a Roman fleet of 200, and lost – but the Romans turned back. For the loss of eighteen ships the Carthaginians foiled the whole Roman initiative. And at Drepana and Lilybaion, the capabilities of a raiding fleet were again shown, repeatedly distracting and annoying the Roman forces, while two whole consular armies – 40,000 soldiers, less casualties – were pinned down in hopeless land sieges. Only when the Romans finally realized that the sieges had to be concluded by the application of sea power was an end in sight. This seems to have been the work of Q. Lutatius Catulus, and even then he could only succeed by raising a fleet by private subscription on the promise of loot. Rome's venture to sea was largely unsuccessful and taught the ruling class little.

Chapter 9

Hannibal's War

Rome's Aegates Islands fleet was the basis for Roman power and authority in the western Mediterranean for the next forty years. The fleet is only attested once in the following two decades, but its existence hovers in the background of the various international crises which the republic faced during that time. In the western Mediterranean it had no competition. Its strength was at least 200 ships. Some of the ships were probably survivors of the 250s, some could be Carthaginian captures of 242/241 (the seventy taken at the Aegates fight), and some could be new, or rather replacement, building. Ships built before 250 are likely to have become rotten by 218, when Rome had 220 in service plus 20 lighter vessels. It is possible, however, that there were more than that. The numbers imply continuous new building. The 200 ships plus the 70 from the 241 victory, became the 220 plus 20 in 218, so some were scrapped, and others will have needed replacing. The replacements were quinqueremes, so it seems, replacing the quadriremes of the Aegates fleet.

Rome was still in expansionist mood during the post-war years. In 238 advantage was taken of Carthage's war with its mercenaries to steal Sardinia, and perhaps Corsica. When these had returned to Carthaginian possession after Rome's conquest in the 250s is not clear, but Rome now decided she wanted them, the result in part of a better appreciation of the geography of the western Mediterranean, and of the possibilities of sea power in the hands of an aggressive Carthage. Already one Carthaginian expedition had gone to the island, though the mercenary soldiers thus transported had immediately deserted. But they so angered the Sardinians that they were driven out of the island – they sailed to Italy for refuge, where they worked to persuade Romans to pay attention.[1] Carthage, after suppressing the mercenary revolt at home, then made preparations to recover the island. At this point Rome began an expedition of its own to the island, and declared war on Carthage when that city objected and continued its own preparations. Carthage, quite unable to face a new Roman war, caved in at once. The city's fleet at the time is said, rather surprisingly, to have consisted of just triremes and penteconters;[2] this seems very unlikely, but the Roman fleet of 200 ships clearly made Rome the master of the situation, and indeed of the whole of the western Mediterranean basin.[3]

The fleet was used in the Illyrian wars in 229 and 219, and in the former case the ships were specifically 200 in number.[4] Hints exist that they may have been used – or some of them, at least – to escort forces to Sardinia and Corsica in the wars of conquest in those islands which followed Carthage's cession; similarly some ships were possibly employed in the Ligurian wars; but there was never any need to send more than a few, if any, on these campaigns. Massalia, the possessor of a small fleet of light ships, upgraded its friendship with Rome to a formal alliance in this time, partly because of their mutual enmity to the resisting Ligurians. This extended Rome's reach to the borders of Spain, where Massalia maintained a trading post at Emporion. In 226 a Roman embassy made an agreement with the Carthaginian viceroy in Spain, Hasdrubal, who had been busy conquering – this was the Carthaginian reaction to the Roman insult over Sardinia. The Ebro River was agreed as a boundary between their spheres, though neither power was anywhere near it at the time. Much of the coast from the mouth of the Ebro to Rome's Ligurian territories was in fact controlled by Massalia.[5]

Roman sea power was thus in large measure latent in those two decades, used offensively only in the Illyrian wars. When the Hannibalic War began in 218, therefore, there was no direct competitor for control of the seas west of Greece. Neither Massalia nor the Illyrians could compete, and King Hieron of Syracuse, who had a small fleet, was another Roman ally. Given Syracusan naval history, Hieron's restraint is another tribute to Roman power at sea. On the other hand, Carthage had been building up its strength in ships, presumably in anticipation of a new conflict with Rome.

The evidence for this has to be extracted from a series of incidents in which Carthaginian ships were employed at the beginning of Hannibal's war. At that time Carthage deployed fifty-five ships to Sicilian waters,[6] and had fifty-seven in Spain.[7] These are no doubt minimum figures, and it is likely that there were more ships at Carthage, since to send out the whole fleet was foolhardy, and perhaps not even possible – eighteen of the Spanish ships were not manned, so that may have been the same elsewhere. On the other hand, as soon as the war began, Rome was able to send 172 warships to Sicily and 60 to Spain;[8] in Sicily Hieron had a squadron of at least twelve ships, which he used to capture three of the Carthaginian quinqueremes which were blown through the Strait.[9]

The naval balance was thus tilted decisively in Rome's favour from the very start of the war, and Carthage made no attempt to alter this, nor did the balance ever change. So this war, unlike the first, was devoid of great sea battles, since it was clear that Rome could bring overwhelming force to bear in any campaign. The Carthaginians accepted that Rome's navy dominated the western Mediterranean. This was one of the bases of the decision made by Carthage and Hannibal to conduct the war by land, though this did not leave Carthage helpless at sea.

Rome's actions at the outbreak of war could perhaps be predicted, but Carthage's took Rome by surprise. One consul, T. Sempronius Longus, took 160 ships, transports, and 26,000 or 27,000 soldiers to Sicily, with the intention of invading Africa. This was intended to be Rome's main theatre of war, a decision so obvious that there was no need for secrecy. The other consul, P. Cornelius Scipio, took a slightly smaller force and sixty ships to invade the Carthaginian province in Spain. Sempronius was delayed in reaching Sicily by the need to suppress a Gallic rebellion in North Italy, and Carthage was able to send two naval forces to Sicilian waters before his arrival. Twenty quinqueremes were sent to the north of the island to raid the Lipari Islands, and perhaps to seize a base there; this force was scattered by a storm. Three of the ships were blown through the Strait and these were the ones taken by twelve of King Hieron's ships. They were taken into Messana. The second squadron, thirty-five ships, went to the Aegates Islands with the intention of seizing Lilybaion by a surprise attack.

The Carthaginian squadron was detected as it approached at night, so the Roman garrison in Lilybaion was alerted. Seeing this the Carthaginians backed off, effectively inviting the Romans to come out and fight. This the Romans did. The size of the Roman squadron is not known, but was certainly less than fifty ships, since it was later reinforced up to that total. Also, exactly where the ships came from is unclear. Hieron, after capturing the three ships in the Strait, discovered the Carthaginian plan and sent messages alerting the Roman praetor at Lilybaion, M. Aemilius Lepidus, and the naval allies. Presumably the latter were the coastal cities in Sicily obligated by treaty to supply ships and men at Roman behest, and it is possible they sent ships to Lilybaion. But it is explicitly Romans who came out to fight at Lilybaion, so either Sempronius had sent some of his ships to Sicily in advance while he was delayed, or there was a permanent squadron stationed at Lilybaion. In view of the outbreak of war, which had been signalled for some time, the latter seems the most likely. (This conclusion implies that Rome had more than 220 ships available, of course – and there may well have been other squadrons here and there.)

The Carthaginians apparently hoped to tap into residual pro-Carthage feelings among the people of western Sicily by their proposed landing, and this was something Hieron was alert to. But the Carthaginian ships are said to have been carrying few soldiers, so it is difficult to see how they could hope to hold Lilybaion even if they had taken it. Relying on a rising by the population would be hazardous. In the event the question did not arise.

The fight outside Lilybaion resulted in a Roman victory, seven of the Carthaginian ships being captured, with one Roman ship holed but saved. The manoeuvring skills of the Carthaginians are emphasized – hence the holed ship, the result of ramming – but it would seem that they were fully equalled by the skills of the Romans, whose chosen method was still to come close and board.[10]

When Sempronius and his army and fleet finally arrived in the autumn, he took the fleet on a preliminary cruise round Sicily, first to Lilybaion, where most of the ships were left, then to remove the Carthaginian garrison from Malta, then to the Lipari Islands to check on whether the first Carthaginian squadron had succeeded in planting a base there, then back to Lilybaion. He missed the active Carthaginian squadron, which had gone on to raid the Italian coast near Vibo.[11]

The Carthaginian actions against Sicily show an intention to revive their old ambitions to rule in the island. Lipari and Lilybaion had been important bases in the previous war, and if Hieron of Syracuse was conscious of sentiment in Sicily in Carthage's favour, no doubt it was also appreciated in Carthage. Had the squadron succeeded in seizing Lilybaion and stimulating a rising, Sempronius would have had to cope with it before attempting a landing in Africa. The prospect of having to besiege Lilybaion again must have scared the Romans, which is why Lepidus was reinforced. One wonders if he was given some soldiers as well as more ships – but there were normally forty marines on each quinquereme, and this would help. Already, therefore, even before the news of Hannibal's exploits had arrived in Sicily, the Carthaginians had made a serious attempt to disrupt Roman plans.

It was late in the year by the time Sempronius returned to Lilybaion, probably October or even early November, too late to risk putting to sea with the whole army. Sempronius was an energetic man, however, as his activity so far had shown, and he might well have taken the risk, but he was recalled by the Senate because of Hannibal's invasion of northern Italy. He set his soldiers marching back northwards (they were probably either still in Italy or had only just crossed into Sicily). Of his 160 quinqueremes and 12 light vessels, he left some with the praetor Lepidus, so bringing his Lilybaion squadron up to fifty ships, and sent 25 to patrol the Italian coast to deter further Carthaginian raids.

He is said to have sailed to Ariminum, the designated muster point for his army, with ten ships, but he is also said to have marched through Rome on his way there, so heartening the people. Perhaps he sent ten ships to the Adriatic but went himself to Rome. The rest of the fleet – between 100 and 120 ships, depending on how many were left with Lepidus, and if any went to Ariminum – were ordered 'home', which probably means Ostia.[12]

He may have left some troops with Lepidus, and there were certainly some marines on the ships he left in Sicily, but he took virtually all the army away. This was in strong contrast to the reaction of his colleague Scipio when he discovered Hannibal's march. He had sixty warships and a fleet of transports to carry his soldiers and supplies, and was at the mouth of the Rhône, when he heard that Hannibal was crossing the river upstream from him. He marched upstream to intercept the Carthaginians before crossing the river, but Hannibal had beaten him to it and was three days' march away. Scipio decided to return to take command of the defence of northern Italy, where there was another Roman army as large as that he was commanding, but he sent his own army on to Spain under the command of his brother Gnaeus, with most of the ships.[13]

This was, of course, one result of having command of the sea. Scipio had the luxury of choice, and he had the ability to move much faster than Hannibal. His fleet had sailed 430 kilometres in five days from Pisa to near Massalia, whereas the Carthaginian army could probably move at less than 20 kilometres a day through the mountains. Scipio returned to Pisa and then rode north through the Apennines, taking command in North Italy long before Hannibal got his army through the Alps. And meanwhile Scipio's own army sailed on to Spain, first to Massalia's trading port at Emporion, and then on to Tarraco, overrunning the coastal settlements on the way; it was paced by the fleet along the coast as it campaigned.

Given Rome's clear predominance at sea, Carthage could only use its own fleet to interfere with Rome's operations by mounting raids, such as those to Vibo and the Lipari Islands, interrupting supplies, picking off isolated ships and small squadrons, using the wider spaces of the sea where the Romans did not go. It was necessary to avoid fighting the superior Roman fleet. The Roman response should be, so the Carthaginians might calculate, to guard convoys with warships, send out searching fleets, and perhaps keep their convoys in harbour longer than intended. This would tend to wear out Roman strength and cause the Roman fleet to waste a lot of effort. The prospect of raids certainly compelled the Romans to use up a substantial proportion of their soldiers in coastal garrisons – Tarentum and other places were reinforced after Hannibal's victory at the Trebbia, partly in response to the raids to Lipari and Vibo.

The Roman maritime supply lines were tempting targets for a raiding force. This meant that paradoxically the maritime initiative lay to some extent with the Carthaginians. A good example came in the next year. A convoy of Roman supplies in merchant ships was sent from Ostia towards Spain, but was intercepted by a Carthaginian naval force off Cosa in Etruria (and this at a time when Hannibal was not far off, in the Trasimene campaign). When the news reached Rome of the convoy raid, a consul was directed to enlist all the available seamen, free, freed, or slave, in Ostia and Rome, and go after the raiders.[14] It was hopeless, of course, for the raiders were long gone even before the news reached Rome, but the loss of supplies was compounded by the disruption caused at Ostia and in the Tiber by the Senate's near-panic reaction.

Such tactics, however, no matter how annoying to the Romans, were not going to win the war for Carthage. That had to be done by the army on land, and that meant Hannibal's army in Italy. But that army, even with reinforcements collected from the Gauls on his way into Italy, was a wasting force. Even if it won all the battles, the army would end up smaller each time, and would need reinforcements to be sent from Africa or Spain, or both. So the raids and the *guerre de course* were secondary to the need to support Hannibal. This the Carthaginians found very difficult to accomplish, and Hannibal had to spend much effort in fighting to acquire a seaport through which he could communicate with Carthage; meanwhile the government in the city had only a vague idea of where he was and what he was doing.

By contrast there are the activities of Cn. Scipio in Spain in 217. He used his fleet of sixty quinqueremes to assist in his campaign along the Catalonian coast in 218, though at one point the fleet had been surprised when beached by an enemy raid.[15] Next year Hasdrubal brought up his own fleet to the mouth of the Ebro. His intentions are unclear, but he had an army with him so his aim was probably to reverse the recent Roman gains. Both commanders had trouble manning their ships. In 218 Hasdrubal had fifty quinqueremes, but only thirty-two were properly manned; he also had two quadriremes, not manned, and five triremes. The fleet he brought up to the Ebro mouth was forty strong, so he seems to have found more men, perhaps from the army. Presumably it now consisted of thirty-five quinqueremes and the five triremes.[16]

Scipio had come to Spain with most of the fleet allocated to his brother. There had been sixty quinqueremes in the original force, but in 217 Cn. Scipio was able to use only thirty-five,[17] which suggests that his brother had taken a few ships, perhaps five, back to Italy with him; he may also have lost some, but Hasdrubal's attack on the straying Roman seamen the year before may have left Scipio with fewer sailors than he needed. He may also have been under the impression that Hasdrubal was unable or unwilling to use the ships he had. He did have help from some small craft of Massalia, who were especially useful since they knew the coast.

Scipio came south from Tarraco with his fleet. He seems to have had information as to Hasdrubal's fleet's position, for Hasdrubal was surprised by the Roman approach. Scipio had sent two Massaliot scout ships ahead, and these located the Carthaginians precisely – though their appearance should have alerted the Carthaginians. Watchmen on the hills and in the watchtowers did raise the alarm, but the Carthaginian sailors were scattered and took time to assemble and man the ships. The Roman fleet was on them before the Carthaginian ships were organized, arriving in order of battle, while the Carthaginians were scrambling to get to sea. The Carthaginian ships turned to get away, but were caught at the mouth of the river, which was too narrow for more than a few ships at a time to get through. Those who were able to do so, escaped upstream, but many of the rest ran their ships on shore and their crews escaped onto the land. Some of the Carthaginians were sunk, but the Romans captured twenty-five ships, partly in the first attack, but mainly by towing the beached ships back out to sea.[18]

The news of this action had repercussions at both Rome and Carthage. In Rome it was decided that the Roman forces in Spain should be reinforced, with both soldiers and ships, and that P. Scipio should go along to exercise joint command with his brother as proconsul. P. Scipio had been wounded in the Italian fighting the year before, but he had now had time to recover, and he had presumably been an advocate for the Spanish war, while Roman instincts would normally be to bring all troops home to cope with Hannibal, who had won two crushing victories on Italian soil. The decision to support the Spanish expedition was a fairly belated one, however, and if Cn. Scipio had lost the fight at the Ebro it seems very likely that the surviving Roman forces would have been withdrawn. His victory now clearly made it worthwhile to go on, if only to prevent Hasdrubal from following in Hannibal's Alpine footsteps with another army. It was as obvious to the Romans as to Hannibal and the Carthaginians that Hannibal's army would waste away if he was not reinforced and supported regularly.

At Carthage, the city's available fleet of seventy ships was sent out to try to contact Hannibal in Etruria – this seems to have been a different fleet from that noted earlier at Cosa, for it sailed by way of Sardinia and then to Pisa. A Roman fleet of 120 ships – no doubt Sempronius' old fleet – set out from Ostia to intercept the Carthaginians, but failed to find it, and then spent some time raiding in Africa instead. This fleet was then left at Lilybaion at the end of the cruise.[19] So twice now a Carthaginian naval force had reached Etruria and got away without the Roman ships doing more than search the seas fruitlessly for it.

The crucial maritime area, however, was in the seas around Sicily. It was between Sicily and Africa that the sea war would be decided, for whoever controlled those cities had access to enemy territory. The major Carthaginian naval base was, of course, Carthage itself; the major Roman base in the area was Lilybaion, the former Carthaginian city. In Africa no place other than Carthage seems to have been employed in this way, but in Sicily Syracuse was the main population centre and a major shipbuilding city, and King Hieron had a small navy. Beyond Sicily, the South Italian cities of Magna Graecia were early targets for Hannibal since they would give him access to the sea and so the ability to make contact with Carthage. A subsidiary element in all this was that there was a proportion of the Sicilian population which was nostalgic for Carthage's rule, just as some of the Sardinians were, if not actively pro-Carthage, certainly anti-Roman. Thus there were plenty of insurrectionary opportunities for Carthage to exploit, and these could be cultivated by sea power. But to exploit them properly Carthage had to remove Roman sea power, at least temporarily, and had to show itself willing to risk its own troops.

The Roman fleet at Lilybaion was used in 217 to raid Africa again, though this time the landing party was ambushed and driven back to the shore. It was taken on another fruitless cruise around Sardinia and Corsica, where hostages were taken to deter anti-Roman actions.[20] Next year the Carthaginians attempted a trap. One part of their fleet conducted a raid into Syracusan territory; a second section lay in wait amid the Aegates Islands for the Romans in Lilybaion to sail out in response. The praetor T. Otacilius Crassus, however, discovered the ambush and failed to move, though he was pinned down for a time.[21] Meanwhile, two separate Carthaginian forces were sent to Spain, with orders to Hasdrubal to take his army to Italy. In the event he was defeated near the Ebro by the Scipio brothers; but in the process he had been reinforced by more ships.

Carthaginian attention, therefore, was largely directed at contacting and reinforcing Hannibal. Hasdrubal pointed out that if he marched to Italy Carthage would probably lose Spain, but he was told to go anyway. But for his defeat at the land battle of the Ebro he would probably have arrived in Italy late in 216. Hannibal gained control of two of the ports in the toe of Italy, Lokroi and Kroton, both crucial for his communications with Africa, and with Sicily, and next year Carthage was at last able to land troops in Italy when a fleet under Bomilcar put forces ashore at Lokroi.[22] The difficulty was that there was a steady Roman maritime patrol along that coast by a squadron of twenty-five quinqueremes commanded by P. Valerius Laevinus. He intercepted a party of envoys between Hannibal and King Philip of Macedon during 215, but he missed Bomilcar's convoy.[23] A second Carthaginian fleet sailed for Sardinia, where the Romans were fighting a local rebellion (despite their holding hostages from the island). The fleet was driven to the Balearics by a storm, and when it reached Sardinia, the rebels had been beaten; they revived, but then lost again. The Carthaginian fleet got away, only to be intercepted by Otacilius' fleet from Lilybaion, which had been raiding Africa. Very revealingly, the Carthaginians did their best to avoid a fight, though they lost a few ships.[24] It seems clear that it was now settled Carthaginian maritime policy to avoid the Roman fleet whenever possible.

This avoidance policy makes sense if raiding and moving supplies was to be Carthage's main maritime purpose, but it did leave Roman sea power intact. The Romans do not seem to have moved in anything less than squadron groups, at least at this period, so the Carthaginians did not have much hope of picking off isolated Roman ships. However, there are no Roman complaints of their supply routes being interrupted, so it seems that Carthage was also not making any attempt to raid the Roman supply lines, such as that which led along the Italian and Gallic coast from Rome to Spain. The reason is probably a lack of bases close enough, but they were able to use the Balearics more than once during the war, and the Sardinian routes from Italy were surely within their reach.

The threat of intervention from Macedon, indicated by the treaty between Hannibal and Philip, sent Laevinus and his fleet, which had been reinforced to fifty ships, from Brundisium across the Adriatic to relieve the city of Apollonia, which was under siege by Philip. Had he gained possession the king would have a worthwhile base from which to reach Italy. In the treaty he had promised to come across with 200 ships, though it is unlikely he had that many, and they were certainly not quinqueremes.[25] (The war in Greece which resulted from this will be considered in the next chapter.) Sicily also became an active theatre of the war in this year. Hieron had died in 215, and the city became politically much disturbed, and veered towards the Carthaginian cause.

It may well have been this combination of threats and problems which persuaded the Senate to order the building of a hundred new ships in 214/213. There had been no new building since the war began, and there were several substantial squadrons stationed well away from Rome by now – thirty-five, or perhaps fifty-five, ships in Spain, fifty at Brundisium – and some of the 220 ships originally available in 218 were at the end of their lives. Meanwhile the war had widened to Greece, there was trouble in Sardinia, and Sicily was disturbed. As a result of detachments, Otacilius' squadron at Lilybaion had fallen to only 30 ships and was now to be reinforced to 100 and later to 130.[26] This fleet was used during the year in eastern Sicily; some ships and a garrison were left at Lilybaion.[27] The total Roman fleet had thus grown to about 280 ships, including a strategic reserve held at Ostia.

Carthage had not fielded a fleet larger than seventy ships until this point, though our sources – almost entirely Livy and Polybios – rarely put numbers to Carthaginian seaborne forces. In 214/213, however, the political crisis in Syracuse provided Carthage with an opportunity to gain control of Sicily, and a major effort was made. Syracuse was the key to the island, now that Lilybaion was firmly held by a Roman garrison. During 214 a series of coups in the city removed the monarchy and brought it to an alliance with Carthage. The Romans assembled a large army, put one of their best generals, M. Claudius Marcellus, in command, and were prepared to spend as long as necessary to take the city, which turned out to be two years.

In addition the Roman fleet was reinforced to 130 ships, some of which lay off Syracuse during the political crisis in the city, though this merely revealed local hostility when a Roman landing was feared and the people rushed to the shore to stop it.[28] Once the siege began, sixty ships were used to attempt an assault from the sea, but were defeated in part by the ingenious defensive measures of Archimedes.[29] The Carthaginians took the opportunity of the Roman fleet being divided between Syracuse and Lilybaion to land an army at Herakleia Minoa on the south coast, in the region which had been Carthaginian even during Agathokles' reign. The Romans landed troops of their own at Panormos, presumably having sailed directly from Ostia rather than coasting, which would have brought them to Messana. Carthage sent a fleet of fifty-five ships to help at Syracuse,[30] but it remained there only briefly, being heavily outnumbered.

Bomilcar returned after the winter with a fleet large enough to deter a Roman attack. He now had ninety ships at Syracuse, but he still was unable to affect the land fighting and could not prevent Marcellus capturing the key Syracusan fortress at Epipolai. Bomilcar took thirty-five of his ships out of the harbour while the Roman blockading ships were preoccupied with a storm, and sailed back to Carthage to report on Syracuse's desperate situation.[31] He returned with a large force, and probably some troops.[32] Disease was ravaging both armies, and the city was also badly affected by famine. Bomilcar went off again, and returned with an even bigger fleet – 130 ships – escorting a transport fleet of 700 ships filled with supplies. He reached Cape Pachynon but was unable to round the cape because of an easterly wind. A Roman fleet came south from the blockade of Syracuse and took station north of the cape. By the time the wind died away, it was too late for the Carthaginian fleet to succeed in its purpose: the Roman force was too powerful to be challenged. Its exact size is not known, but 130 ships had been assigned to Sicily, and probably 100 were on duty at Syracuse. Bomilcar decided that the odds were too great; he sent the transports back to Carthage and took at least some of his warships on a cruise to Tarentum.[33]

This was, in effect, the end for Syracuse, though the actual conquest and surrender took some time yet. A seaborne landing on Ortygia took that fortress and convinced even the diehards in the city that all was lost.[34] The Carthaginian forces in the interior of the island were then driven out or defeated. By the end of 210 all Sicily had been recovered by Rome. This was Carthage's last chance to win the war, as it proved, though this was not realized until some time later.

Tarentum had been taken by Hannibal while Syracuse was under siege, hence the visit to the city by Bomilcar and his fleet, a visit which was repeated next year.[35] The acropolis had been kept in Roman control, and the standoff between city and citadel lasted for three years, with both sides running in supplies. All the old Greek cities of Magna Graecia had fallen to the Carthaginians except Rhegion, and it was from there that the Romans had to send in their supplies; the Carthaginians kept a squadron at Tarentum as a deterrent. Livy describes one Roman attempt in 210 when a squadron of twenty ships was used to escort transport convoys from Rhegion towards Tarentum. The squadron consisted of five Roman ships reinforced by fifteen from the naval allies, all under the command of D. Quinctius, 'a low-born man who had proved himself a competent commander'. He had started with just two ships under his command but had persuaded Marcellus and the naval allies to provide the rest. His convoy was, however, intercepted by the Carthaginian squadron out of Tarentum and defeated. Quinctius himself was killed.[36] On another occasion a Carthaginian fleet ran supplies to the city, but there were so many sailors on board that they consumed more than they had brought, and had to leave.[37]

The Carthaginians, in raising their large fleets in 212, appear to have collected all their various forces together. Keeping watch in Africa – there were regular Roman raids from Lilybaion – mounting reply-raids to Sicily and Sardinia, helping Tarentum, and a squadron sent to Greece, all meant that other areas were deprived of Carthaginian naval help. In 209, when P. Cornelius Scipio (the son of the P. Scipio who fought in Spain, who was killed in 211), made an audacious attack on Nova Carthago, he found no Carthaginian naval resistance. He used his own fleet, and launched assaults from both land and sea. When he had captured the city, the booty included sixty-three merchant ships, but only eight (or maybe eighteen) warships.[38] Next year Scipio was able to lay up his own warships, and incorporated the marines and crews into his army; a year later he sent fifty ships to Sardinia, apparently permanently, and presumably for use there, keeping just thirty for his own use. So for several years up to eighty Roman ships had been kept in Spanish waters, while Carthage had only a few there.[39]

Mutual raiding was carried out in Sicilian and African waters between 210 and 207, the Romans raiding Africa, the Carthaginians raiding Sardinia. Eventually, when several of the Roman raids had been unopposed at sea, the Carthaginians sent out a fleet of eighty-three vessels to intercept a raid from Lilybaion against the Cape Bon Peninsula. Laevinus, the Roman commander, had a hundred ships and must have been rather surprised to be challenged in this way after earlier Carthaginian lethargy. He won the subsequent fight, and captured eighteen ships.[40] Next year (207) – this was when Scipio sent fifty ships to Sardinia from the Spanish squadron – Laevinus conducted another raid to Africa, and the Carthaginians made another attempt to stop it. The Carthaginian fleet this time was only seventy ships, and they lost twenty-one, four sunk and seventeen captured, in the fight.[41] (Roman casualties, if any, are mentioned in neither case.)

Carthage was apparently making no attempt to build new ships, and with the city's fleet now reduced to less than fifty vessels there was little or no hope of reviving its sea power if no new ships were built. Livy comments that, after Laevinus' second victory, 'the sea was safe', and later in the year the Senate reduced the Lilybaion squadron to just thirty ships.[42] This means it was fully understood at Rome, and in Sicily, that there was no danger from the Carthaginian ships in Africa. But it was just at this time that the naval war in Spain revived; presumably this was the result of ships being sent from Carthage, though it is clear there were some at Gades as well.

By this time the Carthaginians had been driven from most of Spain. Only a few places in the south remained to them after their defeat at the Battle of Ilipa. The Carthaginian commanders in that battle escaped the Roman pursuit on ships sent from Gades to collect them.[43] The commander Hasdrubal Gisgo sailed from Carthage with two quinqueremes and seven triremes, and some other ships were used to rescue Mago, Hannibal's brother. Gades clearly harboured a squadron of more then ten ships.[44] A small battle at sea was fought in the Strait of Gibraltar between fleets consisting of one quinquereme and some triremes on each side; the Roman quinquereme was handled well, while the tide caused confusion amongst the lighter triremes. The Carthaginians had been attempting to get to the African coast to reach Carthage, a voyage which was clearly still possible for them.[45] The last Carthaginian commander in Spain, Mago, was ordered to leave Gades, the final Carthaginian post, and go to Italy. He was able to raid Nova Carthago, unsuccessfully, on his voyage, and spent the winter of 206/205 in the Balearics. His casualties at Nova Carthago were nearly 3,000 men, which suggests that he had a considerable fleet, some of which were presumably transports, and next year he reached Genoa with thirty ships.[46] He seems to have had no fear of being intercepted by a Roman fleet. No doubt Scipio's ships were still largely out of use.

By this time Roman attention was devoted to the possibility of ending the war by an invasion of Africa, though many in Rome were unenthusiastic, since the war was essentially won. Scipio, fresh from the conquest of Spain, was elected consul, virtually by popular acclamation. He was given permission to cross to Africa, but was not given enough troops needed for the enterprise. Volunteers enlisted, however, and he gathered a fleet of thirty ships to add to the thirty which were in the Lilybaion squadron. Carthage replied to this by diplomatic efforts to gain allies, and by reinforcing Mago, who now had reached northern Italy, landing at Genoa. He had used thirty ships in his voyage, of which he kept ten and sent the rest to Carthage; Carthage then sent him twenty-five in return, with armed help, along with instructions to recruit an army among the Ligurians and Gauls. A Carthaginian fleet of eighty transports, which was captured off Sardinia, was presumably heading for North Italy as well.[47] Mago did his best, but was unable to penetrate into Roman Italy, still less to force the withdrawal of Scipio from Africa. Carthage itself was clearly unable to build new ships – lack of will to fight is not the issue, for when Scipio did invade he faced stiff resistance. Perhaps the lack of resources (if that is the reason for the lack of ships) was a result of Roman control of the sea, and of the repeated raids on the African coast.

Scipio clearly had no fear of any Carthaginian naval force when he sailed from Africa with his army, in a fleet of 400 transports, escorted by only 40 warships.[48] He evidently left twenty ships at Lilybaion as a reserve and as a protection for Sicily. Next year, as he campaigned in Africa, the Senate reinforced that Lilybaion fleet to forty ships, sending thirteen new ships along with some older ones. The Senate was also clearly uncertain that the Carthaginian naval threat had really ended: 160 ships were in commission in 203, 40 each in Sicily, Sardinia, and guarding the Italian coast. (The other forty are not specified; South Italy seems a likely posting).[49] It was no doubt some ships of the Sardinian squadron which interrupted the convoy carrying Mago's troops back to Carthage after he and Hannibal were recalled by the city government. In both cases, however, most of their troops reached Africa – but in transports, not, it seems, in warships. Their arrival did not, therefore, increase Carthaginian naval strength.[50]

The source material for understanding ancient ships consists of written descriptions (of usually considerable obscurity), some paintings, mosaics, sculptures (particularly bas-reliefs) and drawings, usually graffiti. None of these is necessarily accurate, since few artists were familiar with the sea. The above two mosaics shown here are from Tunisia: the story of Odysseus listening to the Sirens (top), and a sea fight (bottom). In both cases we can be sure that the mosaicists have attempted to show warships. It will be seen that the problem which preoccupies modern researchers, the layout of the oarage, is just that part which the artists chose to indicate without detail. However, the likely variety of sizes, shapes, bows and sterns, with or without sails, are suggested.

Battleship *Isis*: The dark fresco (top) from Nymphaion in the Crimea does not photograph well, but it gives an indication of the ship. The line drawing (bottom) brings out much of the detail. As with so many ancient ship illustrations, the oarage is skimped though a valiant attempt has been made to show most of the oar ports as small holes. The artist shows the ship as though suspended in air, but does thereby show the whole ship, and by omitting people he can avoid the problem of scale distortion. (The eagle flying above and the elephant above are not connected to the ship.) The name *Isis* can just be seen on the drawing at the bow. The size of the vessel is indicated by its two steering oars – it was probably a quinquereme. Despite its limitations (no mast is shown, for example) this is undoubtedly one of the better illustrations of a ship from the ancient world.

(Above) A Roman trireme on a carved panel from Puteoli (Pozzuoli) on the Bay of Naples. The crew are shown out of proportion to the ship and the oars are mainly indications, though the ends indicate that there were two banks. The size is thus an assumption rather than anything directly attested. The man in the stern is the steersman, and the man in front of him, shown slightly larger than the soldiers, is probably the trierarch. The date is the first century BC.

(Left) Ship construction. The method of building ships was to form an outer skin as the hull, then bind it together with internal ribs. The result was flexible, useful in the short, steep seas which can develop in the Mediterranean. The deck on larger ships helped bring the whole together – though many smaller ships were open (*aphract*). This example is of a Roman vessel excavated from the mud of the Rhine at Mainz and is displayed in the Ship Museum there, but it is typical of ship construction throughout the ancient world.

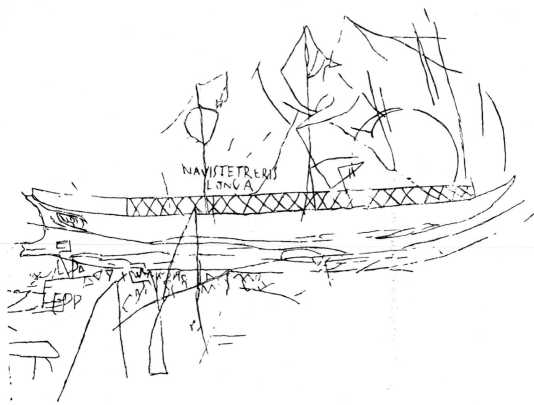

'*Navis tetreris longa*'. The only ancient drawing of a ship with a label indicating its type – a *tetreris* or quadrireme. It comes from, of all places, Alba Fucens in the Apennines, well away from the sea, and is probably of the first century BC. This makes it a Roman ship of the type used at Actium. It is lightly built without bulwarks. The artist indicates sails roughly but omits the oars. The sea is also indicated impressionistically. The latticing along the hull would provide ventilation for the rowers.

The downstream end of the Isola Tiberiana (Tiber Island) in Rome was carved and built into a stone sculpture of a quinquereme, suggesting Rome was somewhat proud of its navy. This drawing shows what the original ship which modelled for the sculpture would have looked like. The oarsmen are omitted, but the modern artist can show a few crewmen at the correct proportion.

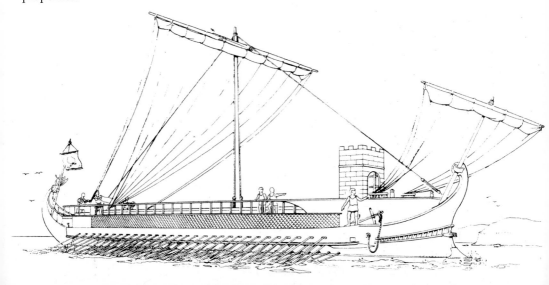

Rhodian galley. A relief at Lindos on the island of Rhodes showed a ship, probably a quadrireme, which was the vessel of choice in the Rhodian navy. It is very faint now and incomplete but this drawing suggests what it would have been like originally. The surviving part shows the stern of the vessel with just an indication of the steering oar (which has broken off) and the high, elaborate stern, whose shape and height helped the steersman and at the same time acted as a balance for the metal ram on the bow.

A mosaic from the Piazza Barberini in Rome shows a fully manned and armed quinquereme going into action, mast down and deck crowded with soldiers. The scene is set in Egypt (with a fishing boat made of papyrus at top left). This is not to be taken as an accurate representation, but as an impression of speed and power it is well done.

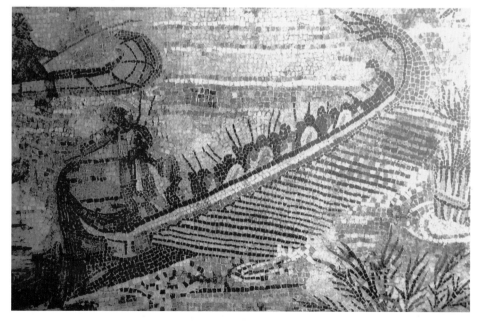

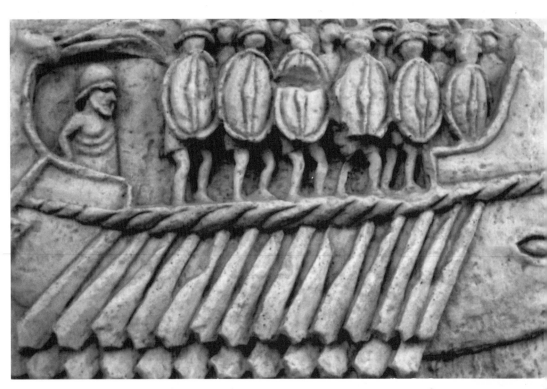

These reliefs show the difficulty artists had in depicting men on warships. The men had to be drawn to a size which showed what they were doing, which meant they were wholly out of proportion to the ship. In the relief from Naples (top), the steersman could not possibly see where he was going, and the number of oars is scarcely enough for a warship. But as a menacing indication of a warship on the attack it is crude but effective. The second relief (bottom), from Rome, is rather better carved but the artist has the same problem and shows only a few of the soldiers, though they are even more menacing. The ship is thought to be bireme, which word Livy uses for *lemboi*.

Antiochos III. One of the few men who feature in this account of whom we have a reliable likeness, Antiochos was one of the great warriors of his day. On his last campaign of the Fifth Syrian War, in which he had already conquered southern Syria, he set out from Seleukia-in-Pieria and captured Korakesion. Later of course, he was defeated at sea and on land by the Roman Republic.

(Bottom left) Demetrios Poliorketes was defiant after the defeat and death of his father at Ipsos (301 BC). This is one of the coins he issued in the years after, showing the winged goddess Nike blowing a trumpet to signal close action, while standing on the stem of a galley, its ram pointed at the enemy.

(Bottom right) Demetrius' grandson, Antigonos Doson, was less aggressive but issued this coin with a relaxed Apollo on a galley, perhaps indicating confidence in his power at sea.

Wherever there was a gently shelving beach it was possible for the crew of a ship to go ashore, perhaps to land its goods, or at least to rest overnight. This beach is at the smally city of Olympos in Lykia, which was built a little inland and on the hills. The city was close to the everlasting fire of the Chimaera and so profited from the tourist trade, but the other factor which kept a small city away from the coast was that such beaches as this could allow pirates to land and raid.

Where a beach was not available, as along much of the coast of Syria, which is often rocky, inlets could be used as ports. This is ancient Gabala (now Jeble), north of Lattakia, Syria. The harbour consists of two parts, as with much greater ports – an inner harbour (on the right), which is rectangular and was artificially enlarged, and an outer harbour which is wider and protected by a breakwater, the modern version of which lies on ancient foundations. This was a small town but its harbour was large enough to hold a number of commercial vessels.

Bay of Haifa. Haifa is now a large modern city, but its main predecessor in the bay was Ake, later developed by the Ptolemies as Ptolemais (and now the city of Acre). It was a major administrative centre and naval base during the Ptolemaic period of control and later. The foreground shows modern Haifa, with Acre at the far end of the built-up area in the middle distance. The bay was the main reason for both cities, being wide and yet sheltered.

The Seleukids in Syria had no ready-made ports in their short stretch of coast, apart from small places such as Gabala. Seleukos I founded two new port-cities, Seleukia-in-Pieria and Laodikeia-ad-Mare (Lattakia), and at each had an artificial harbour excavated. That at Lattakia still operates as Syria's main port; that at Seleukia (shown) has since silted up, returning to the condition before the city's founding. This photo is taken from a hill within the ancient city, looking out over the old harbour. The coast is just visible on the right. On the left, beyond the buildings, a darker area with a curved hedge is the site of the original circular harbour; an outer harbour lay between it and the sea, roughly where another lush area is bounded by two straight hedges.

Korakesion (Alanya), Pamphylia. The rock of Korakesion stuck out into the sea, forming sheltered bays either side. The first view (top) is from the hill looking over the modern harbour, which was also the ancient and medieval harbour. The second view (bottom) is from the harbour towards the hill and shows how steep the climb is. At the foot of the hill are ship sheds (the line of arched openings), now of medieval date, but maybe originally Hellenistic. The fortifications on the hilltop are certainly originally Ptolemaic and the city was a Ptolemaic fortress which defied the victorious Antiochos III for a month. Behind the harbour (top) are the hills of Rough Kilikia, the home or haunt of many of the pirates who so bothered Rome. Pompeius also had to besiege and capture Korakesion.

(Left) One of the old Greek cities brought within the Ptolemaic empire was Phaselis in Lykia, which had long had active commercial relations with Egypt – it was probably a source of wood for treeless Egypt and so important for the Ptolemaic navy. Considerable construction took place in the city in the Ptolemaic period, including the quay in the foreground of this view of the north harbour. The rocks in the middle distance are the remains of breakwaters, also of the 3rd century BC, and another was built in the south harbour. The Ptolemaic investment in the city was thus substantial.

(Below) While the Romans besieged Lilybaion and Drepana on the western end of Sicily, Hamilcar Barca harassed them first from Hierkte on the north coast and then from this hill, Eryx (now Erice). The visible fortifications are mainly medieval, but the native strength of the position is obvious and the foundations are ancient. There was also a famous temple of Venus on the site.

Carthage. The two harbours at Punic Carthage can still be seen and visited, though the ancient buildings have now gone. The first view (top) looks from the 'Admiralty Island' in the middle of the circular naval harbour, which held administrative buildings, and across to the long rectangular commercial harbour. The aerial photograph (right) shows the condition of the harbours now. The naval harbour was ringed by ship sheds, the commercial harbour by wharves and warehouses.

With Scipio on land in Africa, and his fleet so relatively small, there was an opportunity for the Carthaginians to use their own ships to score a naval success. When Scipio began besieging Tunis as well as Utica, the Carthaginian fleet came out of Carthage city to attack the Roman ships at Utica. But they were so slow that Scipio was able to ride back to his camp and organize an effective defence before they arrived. His warships were moored in a line, and protected by three lines of merchant ships lashed together in front. When the Carthaginians attacked the fight actually took place on those merchant ships. The Carthaginians managed to seize sixty of them, and regarded this as a success, but none of the Roman warships was touched.

The Carthaginians' move had been sensible, since many of the warships had been used in the siege of Utica and were burdened with siege machines; if those ships were sunk or seized, the siege would have been relieved and the local Roman sea power severely damaged. But the slowness of the Carthaginian fleet, which took two days to sail the fairly short distance from Carthage to Utica, sabotaged their intention. It looks as though the fleet was very much out of practice.[51]

The ships of Carthage were used only once more. Peace terms had been agreed in Italy, when a Roman supply convoy heading for Africa was scattered by the storm and some of the ships were driven into the bay close to the city. A popular clamour insisted that the ships be captured and the supplies used to feed the people of Carthage. This broke a local truce, and as a result the peace talks were broken off.[52] But Hannibal's defeat at Zama later finally sealed the matter, and in the peace terms as finally agreed, all the Carthaginian warships was surrendered; Scipio burned them, to the dismay of the citizens. Livy reports that some of his sources claimed there were 500 of the ships, of all sizes. If so, and Livy clearly had his doubts, many of them were no doubt small and others old.[53] The lack of enterprise by the Carthaginian fleet in the latter years of the war, especially during the war in Africa, suggests a lack of suitable ships, or a lack of manpower, or a lack of resources, or a combination of all three.

Rome's command of the western Mediterranean was never seriously challenged during this war. Even when the Carthaginians had managed to assemble a fleet of 100, then 130, warships in the attempt to relieve the siege of Syracuse, that fleet was faced by a Roman fleet of almost the same size, while other Roman fleets also existed in several places from Greece to Spain. Further, those Roman fleets were manned by experienced oarsmen, sailors, and marines, while the Carthaginian seamen were, at least at the end of the war, in short supply. For some time the Carthaginians had made a reasonable attempt to use their naval resources sensibly, by raids on Roman territory or by intercepting Roman supplies, but the only way to beat the Romans at sea was by sinking their ships, and Carthage never had the strength to do this.

Chapter 10

Philip V

The previous chapter considered the naval aspect of Hannibal's war in western waters; that war also spilled over into Greece when King Philip V of Macedon made an alliance with Hannibal, though it was really a separate war, one which actually had more to do with Greek affairs than Italian. From Rome's point of view Philip was a marginal problem, a nuisance rather than a threat, though he still had to be attended to.

King Philip contacted Hannibal in southern Italy, and the two men made an agreement to cooperate against Rome, though the terms were more on how to divide up the potential imperial spoils than on how to cooperate in the fighting.[1] The Senate obviously had to assume that an invasion of Italy by Philip was intended, so it increased the Roman naval forces in the south and stationed them at Brundisium, rather than at Tarentum. From there the squadron would be able to intercept any invasion force coming from the east;[2] later, in a sensible tit-for-tat, Roman envoys went to Greece and made an alliance with Philip's local enemy, the League of the Aitolians. These measures pinned Philip down in Greece and prevented him from sending reinforcements to Hannibal.

Philip's difficulty was geographical. His main kingdom was separated from the Ionian Sea by the Pindos mountains, and the Illyrian coastal communities were neutral or hostile to him. To reach Italy he needed to establish a secure base on the eastern Adriatic coast. His first move was against that part of the Illyrian coastline which Rome had made into a protectorate a few years before, at the end of the Second Illyrian War.

The Roman naval squadron stationed at Brundisium blocked Philip's moves. In 214 he made an amphibious attack on the city of Apollonia in Illyria, which was a Roman ally (Rome had rescued it from an Illyrian siege in 219/218).[3] Philip was using *lemboi*, a type of ship that had been developed in Illyria. In this he was influenced no doubt by Demetrios of Pharos, who had been a leading figure in Illyria until driven out by Rome in 219 at the end of the Second Illyrian War. He had been given refuge by Philip and persuaded him to use the small ships; Philip at that stage seems to have been very impressionable.

The *lembos*, though it had a name which suggests Illyrian origin, was in fact another version of the light craft which were used in all navies for many of the tasks below the level of fleet action. In early Greek days penteconters had been the normal type; Athens used *triakonters*. These various types were all used as swift and manoeuvrable craft which could go into shallower, narrower waters than the bigger vessels. Polybios, in discussing Philip's *lemboi* explains them well, in words which can be applied to all the light craft: 'to transport troops, to move swiftly where he wanted, to make sudden unexpected appearances before the enemy', but they were not intended 'to match Rome in a formal sea battle'.[4] *Lembos* was a very elastic term which could cover a variety of ship types and sizes.

These are clearly perfectly good vessels for warfare in Illyria, where there were no ships larger than other *lemboi*. The problem was that Rome had no intention of fighting on Philip's terms. He was fully aware of Rome's naval might, and in an earlier expedition to the Ionian Sea in 216 he had abandoned his approach on the mere rumour that a Roman squadron of quinqueremes was approaching, even though it was said to be as far away as Rhegion.[5] But Hannibal's successes in winning battles in Italy in 218–216 suggested that Rome was sufficiently battered and preoccupied that Philip had an opportunity for gains in Illyria. He would have been better advised to do so before making the agreement with Hannibal, since the capture of his envoys had alerted Rome to the problem and caused the reinforcement of the south Italian squadron. And he should have been aware that the defeats of Rome on land had made no difference to its strength at sea.

Philip managed to bring a fleet of 120 *lemboi* to Apollonia, captured Oricum at the head of the gulf of Aulon, south of Apollonia, which was a sort of outpost of the city, and then laid siege to the city itself before the Romans at Brundisium knew that anything was happening. The successive commanders of the Roman forces at Brundisium had, as part of their task, to search out information about Philip's activities.[6] No doubt this was known on the other side of the water, so when he was told, by envoys which had escaped from Oricum, of the Macedonian expedition, the current Roman squadron commander, P. Valerius Laevinus, acted quickly. He took the squadron across to Illyria, a voyage that took only two days, recaptured Oricum, and landed a force which quickly relieved Apollonia. Philip was so surprised that he fled, leaving his clothes behind, his military equipment for the Apollonians to appropriate, 3,000 of his soldiers dead, and on the way he burned his ships.[7]

Philip had been ruling as king since 221, and was ambitious to revive Macedonian power, which had been battered badly in the reigns of his predecessors, his father Demetrios II (239–229) and his stepfather Antigonos III Doson (229–221), who had acted as regent for Philip for his whole reign. They had had to fight most of their reigns, and one of the enemies had been the Ptolemaic kingdom, which was Macedon's neighbour in Thrace and controlled the Hellespont, and in the 220s had financed the resurgence of Spartan power in the Peloponnese. Doson gave up the posts his predecessor had held in Athens in 229, and this clearly lessened the importance of the posts held by Macedon in the western Kyklades, where the Island League seems to have become defunct. Perhaps in response to this Doson launched an expedition right across the Aegean into Karia in 227.[8] This did not last long, but it did challenge the power in that region of both Ptolemy and Rhodes. This expedition took place at about the time Rhodes city was destroyed in a great earthquake; one wonders if there was a connection.

Doson's conquests in Karia were only temporary and left little or no trace, though Philip clearly remembered Macedonian interest in the area. It appears that whatever Doson captured was handed over either to a local dynast, Olympichos, who had started out as a Seleukid governor twenty years before, or to Ptolemy III, who soon after abandoned his earlier support for the Akhaian League, Doson's enemy at the time, in favour of Kleomenes III of Sparta; again, one wonders at the coincidence.

From a naval point of view the period of the reigns of Demetrios II and Antigonos III shows that there was still in 227 a worthwhile Macedonian navy, capable of carrying an expedition right across the Aegean. It was probably an inheritance from the reign of Antigonos Gonatas, so the ships would be getting old by 227. The Karian expedition was in fact the last time that fleet seems to have been used; by the time Doson died, in 221, it had apparently ceased to exist. If the ships were as old as the Gonatas' reign, most of them will certainly have become rotten and unseaworthy by 221.

The decay of the Antigonid fleet relaxed the pressure on the dominant naval powers in the Aegean, Ptolemy IV and Rhodes, and the former's preoccupation with war in Syria between 221 and 217 no doubt reduced his presence still further – most of the available forces were recalled to Egypt to defend the kingdom's heartland, which was under direct threat for the first time in decades. Philip of Macedon's ambitions were thus made more achievable by the hollowing out of Ptolemaic power. During the campaign in Syria, most of the Syrian ports controlled by Ptolemy were captured by Antiochos III, and his Syrian naval squadron was taken. For the first time, therefore, the Seleukid kingdom had a moderately powerful naval presence on the Syrian coast. Philip's ambitions brought him inevitably into conflict with other Greek powers, since he appears to have defined his area of influence and power as Greece itself. Above all he came into conflict with the Aitolian League, and the two fought a war (the 'Social' War) between 220 and 217. Then he went on to develop his naval capacity, and this brought him into conflict also with Rhodes, and with the Ptolemaic presence in the Aegean.

In the generation since the defeat of the Ptolemaic fleet at Andros in 245 Ptolemaic power in the Aegean, and elsewhere, had gradually decayed. The accession of a new Ptolemaic king in 221, Ptolemy IV, opened the way for a revival of the conflict with the Seleukid kingdom over Syria. In the Fourth Syrian War (221–217) Antiochos III conquered all Syria before losing the vital battle at Raphia, a battle in which even more troops were deployed than in any of the Romano-Carthaginian battles of the same period. Antiochos had to withdraw from most of his Syrian conquests, but he did retain two major gains. First, he had recaptured the city and port of Seleukeia-in-Pieria, and second he captured a substantial fraction of the Ptolemaic fleet, forty fully equipped warships.[9]

Until this war the Seleukid kings had ignored the sea. Their empire, which stretched east to the borders of India and west into interior Asia Minor, was continental in size, and with the loss of Seleukeia-in-Pieria, and the Ptolemaic monopoly of the western Asia Minor ports after 245, it had few outlets to the sea, and no opportunity for developing a navy; there is no record of any Seleukid warships or naval ability until this war. They may well have possessed a few ships – there is one man, Diognetos, who was referred to as a sea-commander (*nauarchos*)[10] – but the great and powerful Ptolemaic fleet was so overwhelming that it made using any Seleukid warships pointless, since they would quickly be sunk or captured in a sea fight.

But now Antiochos had the nucleus of a fleet, and the recapture of Seleukeia-in-Pieria gave him a major port in which to keep it. For the next fifteen years Antiochos was constrained by the peace treaty with Ptolemy IV which ended the Fourth Syrian War, but during that time he concentrated on shoring up his kingdom in the east and extending it in Asia Minor. Meanwhile the Ptolemaic fleet had been seriously reduced, and does not seem to have been built up again.

Ptolemy IV Philopator, however, did have one ship built. This is the biggest of the monster vessels, a forty, of which a detailed description has survived in one of Athenaios' excerpts.[11] It was a catamaran type, of course, with two great vessels joined by a plank deck, and carried thousands of men as rowers, sailors, and soldiers. It was also virtually useless, clumsy and slow, and essentially confined to the Nile. It was therefore for show, not for use, a grand gesture by an imperial ruler, designed to impress. It certainly did that, but the resources, ingenuity, and time spent on it would have been better used to maintain the fighting fleet, whose major units, if they still existed by this time, dated from the 260s or thereabouts. It would seem that his unexpected victory in the Fourth Syrian War had persuaded Ptolemy, and perhaps his courtiers, that there was nothing to fear from any of their possible enemies. It is evident from later events, however, that the fleet had been considerably reduced in numbers and power, and the loss of forty ships to Antiochos III was a serious blow. It also seems that, during his absence in the east and in Asia Minor, Antiochos was quietly building up a new navy.

The Aegean area was where the several Great Powers came into closest contact with each other: the Seleukids controlled the Asia Minor interior and

Antiochos III was steadily encroaching on independent cities and coming closer to the coast; the Ptolemies held several cities along that coast and the islands near to it; the Antigonid kingdom had Macedon and Thessaly, and so the whole north and northwest coast of the sea, and had interests throughout mainland Greece and the Aegean. After 217 all three of these kingdoms had major navies. In between them there were several lesser powers: the Aitolian and Akhaian Leagues on the Greek mainland, the Attalid kingdom in northwest Asia Minor, and Rhodes in the islands. Of these Aitolia had a greatly exaggerated reputation as a source of pirates, and Rhodes had a navy which had partly filled the maritime space left by the internal decay of the Ptolemaic empire, while the Attalid kingdom was mainly preoccupied with the proximity of, and the growing power of, Antiochos III, but had developed a small and efficient fleet at its port of Elaia.

Rhodes maintained a fleet of about forty ships at sea or immediately available, but was able to put double that to sea in an emergency. Most of the ships were quadriremes or *trihemioliai* which were fast and manoeuvrable, qualities enhanced by the outstanding skill of the captains and rowers. The whole citizen body of the city was involved in the fleet in one way or another, the rich as trierarchs, supplying and captaining their own ships, the poor as rowers and soldiers. It has been suggested that every citizen would probably serve in the fleet more than once in his life. There were numerous onshore private associations devoted to maintaining seaboard solidarity and friendship on shore; on their gravestones Rhodians always emphasized their naval service above everything else.[12]

The city was therefore an impressive naval power, but it was small compared with the navies of the Great Powers, and its range was limited to the Aegean, and the coasts round Karia and Lykia. Its prominence was in large part the result of the fact that after the battle of Andros the Ptolemaic power came to be concentrated along the coast of Asia Minor, and became relatively inactive. The foundation of Rhodes' power was the use it made of its fleet in suppressing piracy, an occupation which brought it the gratitude and friendship of the smaller states, and led it to take the Island League under its wing. Rhodes was the only Aegean state never to compromise, never to cut deals, with pirates. It was also devoted to keeping open the sea lanes, at least for its own merchant ships.[13] It fell into a war with Byzantion in 220 because that city insisted on levying tolls on ships passing through the Bosporos, from the proceeds of which it paid the tribute required by a voracious and threatening Celtic kingdom inland. This struck Rhodian nerves in two registers: by restricting the freedom of passage of ships and by paying political blackmail.[14]

At about the same time, after 220, the arrival of Philip V to power in Macedon began the revival of Macedonian sea power, under the influence of Demetrios of Pharos.[15] But Philip was impatient. Not only did he settle for the smaller craft, *lemboi*, recommended by Demetrios, which could be built relatively quickly and cheaply, but he also had influence in Crete, a notorious source of both mercenaries and pirates, where he had been elected as president of the island confederacy (an honorary but influential post).[16] Eventually he was to hire some of the pirates in Crete to operate as his allies. His proximity to Rhodes also grated on Rhodian nerves from the start; his eventual use of Cretan pirates brought hostility.

The Attalid kingdom had gained access to the sea at Elaia, on the coast of Asia Minor, but was mainly based inland at the fortress city of Pergamon. The kings had pretensions to be the saviours of the Greeks in Asia Minor and to be cultural leaders, and during the civil wars in the Seleukid kingdom in the 220s Attalos I had briefly conquered much of Asia Minor. He did not keep it, but held on to his basic kingdom, and soon extended his influence north to the cities of the Straits. By about 210 he had a fleet of at least 35 quinqueremes which had presumably been built up gradually over a period.[17] This, in the Aegean context, made Attalos a player of some note in the international game.

Rome's arrival on the naval and political scene in the Aegean therefore added one more item to an already complex and volatile situation. Philip, in default of a useful naval counter to Rome's power in the Adriatic, made serious inroads into Illyria by campaigning overland from Macedon. Philip captured the town of Lissos, north of the Roman part of Illyria, so gaining an Adriatic port; if he wished he could now reach Italy.[18] This was the point at which Rome contacted the Aitolian League, which had serious grievances of its own against Philip from their last war. Syracuse and Capua were still under Roman siege, and Hannibal took Tarentum. If ever there was a moment when it looked as though one more push would bring Rome down it was then.

And yet such a fraught time was not the best for negotiating an alliance. The Aitolians could see perfectly well what Rome wanted, and they were not interested in doing Rome's fighting, though gaining revenge on Philip and cutting down Macedonian power was certainly something which attracted them. It took a year to agree a treaty of alliance, by which time a good deal of the Roman urgency had dissipated – Syracuse fell in 212, Capua in 211, both victories releasing substantial Roman forces for other work. In the result the treaty of alliance included a provision that the Romans should provide a fleet of 25 quinqueremes at least; Aitolia would provide the army. Neither side was to make peace without the other.[19]

This, for Rome, was a cheap price to pay for security against Philip's supposed plans against Italy, for it committed the city to providing no more than a tenth of its fleet to the joint cause – while Aitolia had to contribute its full armed strength. For the next few years Aitolia and a group of other Roman allies in Greece fought on land while the Roman fleet operated at various points around the coast. In a useful early show of determination, the fleet seized the island of Zakynthos and the coastal city of Oiniadai soon after the alliance was formed. Both, in accordance with the alliance terms, were handed over to the

Aitolians.[20] Next year, probably at an Aitolian suggestion, the town of Antikyra in the Gulf of Corinth was taken. This was a key point in Philip's line of communications with the Peloponnese and his Akhaian ally. The town was given to the Aitolians; the goods and the enslaved population went to the Romans.[21] The Greeks affected to be shocked at this, ignoring all the many cases when the Greeks had inflicted such unpleasantness on each other – the display of shock thus seems more like propaganda than reality. When Philip attacked Echinos on the Malian Gulf, the Roman fleet sailed round into the Aegean to bring help. The island-city of Aigina was then seized; it was handed to the Aitolians, but the fleet continued to use it as a base.[22]

This pattern was repeated for the following years. Philip was able to move about by land, and Rome's allies came under attack. The Roman fleet was able to raid almost at will, but it was not going to win the war that way. The fleet was as ineffective in Greece as was that of Carthage in the wider war, and it used much the same tactics. But Rome's purpose in the Macedonian war was merely to occupy Philip in such a way as to keep him and his army out of Italy. To Rome this was very much a sideshow. In 209 the fleet raided in the Gulf of Corinth again, and the landing parties were driven away by Philip;[23] it may be that the fleet had earlier helped to convoy an Aitolian army across the mouth of the Gulf, though this is not attested. Philip made an attempt to gather an opposing fleet, getting ships from the Akhaians and perhaps from the king of Bithynia, his father-in-law.[24] A Carthaginian squadron under Bomilcar came across from Tarentum but seems to have got no further than Corcyra, presumably because Bomilcar felt it was not strong enough to challenge the Roman fleet.[25] Its arrival was, however, well enough timed to persuade Philip to break off peace negotiations.

Corcyra was a crucial island which is generally ignored in all this. It had come under Roman protection in the First Illyrian War in 229, when the Romans rescued it from Illyrian occupation. In 213/212 and 212/211 Laevinus' fleet wintered there. It was, that is, the main Roman maritime base on the Greek side of the Adriatic. Based there, the squadron dominated the whole coast from Illyria southwards, and was able to take action in the Gulf of Corinth too. The Carthaginian squadron under Bomilcar from Tarentum went to Corcyra in 209. No action is reported, and Philip clearly hoped it would come into the Gulf, but the Carthaginians presumably could not do anything at Corcyra, and to sail on past, leaving the Roman squadron behind them, was too dangerous. The Carthaginians sailed away, presumably back to Tarentum, which was retaken by the Romans that year. When Corcyra was threatened, the Roman fleet seems to have been unable to move, but the Roman position was strong enough to deter a full Carthaginian attack – no doubt Bomilcar instituted a blockade.

Another ally came to join the Roman-Aitolian alliance, King Attalos of Pergamon, who was bribed by the Aitolians with the island of Aigina, sold to him for thirty talents.[26] He came across from Elaia with thirty-five quinqueremes and he and P. Sulpicius Galba wintered together at Aigina. In the next year they cruised to Lemnos and back to Peparethos before raiding in Euboia and Lokris.[27] Attalos then went home when Prusias of Bithynia attacked his kingdom. The Carthaginian fleet reappeared, either having wintered in Illyria, or coming all the way from Carthage (for Tarentum had been taken by the Romans by this time and all Sicily was now theirs also). Philip gathered another small naval force, three quadriremes and three biremes from the Akhaians, with seven of his own quinqueremes and twenty *lemboi*, but he failed to link up with the Carthaginians, who turned back from Oiniadai when Bomilcar heard that the Roman fleet had left Echinos to return to the west. They had been able to reach the mouth of the Gulf because the Roman squadron was in the Aegean – and they had sailed past Corcyra this time – but Bomilcar could not risk being trapped inside the Gulf.[28]

By this time Bomilcar's nearest base was Carthage, and he had to be cautious, for he was in command of what may have been Carthage's last fleet. But the Carthaginians' very presence pulled the Romans back to the west. Philip was able to conduct raids into Aitolia before Sulpicius with the fleet arrived, and got away by hauling his ships across the Isthmus. As he launched the ships again Sulpicius was still actually at Aigina; Philip simply sailed past him.[29] It had taken some time, but Philip was clearly using his limited naval resources rather well, taking advantage of the fact that the two major naval powers effectively neutralized each other. Inspired by this, he made plans for a hundred-ship fleet, to be built at Kassandreia in Macedon, but once he realized the cost, the plans were cancelled.[30]

The Aitolians could not see an end to the war, and no further Roman assistance than the 25 quinqueremes would come. The legion of soldiers which originally came with the ships had been withdrawn when Laevinus was replaced by Sulpicius, though there were plenty of soldiers in the ships, and a thousand men from the squadron fought in a battle at Lilaia in 208. So in 207, when, for the third time, a group of neutral states offered to mediate, the Aitolians accepted, though Sulpicius demanded that they continue the war.

The mediation which succeeded in 207 was in fact the third attempt. The first was in 209 when representatives of Ptolemy IV, Rhodes, Athens, and Chios met with the Aitolians and Philip near Lamia and persuaded them without too much difficulty to agree to a truce in preparation for a general peace conference. But this conference broke up when Philip heard extravagant Aitolian demands, which were in fact the demands of the Romans, whose squadron lay nearby. It was not in the Roman interest that peace be made.[31] The motives of the mediators were to try to avoid either Rome or Philip becoming too powerful in the Aegean. Athens was by now a professional neutral, but still an important commercial city. Egypt and Rhodes were important naval powers who could see that an increase in the power of Macedon, or the permanent presence of Rome in Aegean, or even Greek, waters, would be a threat to them – this is one of the signs of the fading of Ptolemaic power in the area. Chios was perhaps included to highlight the danger of an increase in power by Attalos, who was a naval neighbour of the island.

This conference failed, thanks to the Romans' influence on the Aitolians. The second attempt, by Ptolemy and Rhodes in 208, was even less successful. The Aitolians refused to negotiate, simply sending the self-proposed mediators on to Philip. Philip was under great pressure at the time, but was nowhere near beaten, and there had been a Roman contingent at the meeting with the Aitolians. This time Athens and Chios are not mentioned as mediators.[32] It has been assumed that Athens at least had a representative there, but there seems no real reason to accept this; it was the two naval powers which were the most concerned, which is the point to note. Once more the possibility of naval competitors was clearly unwelcome, and both Rhodes and Ptolemy knew perfectly well that it was such wars which stimulated powers to build ships; it was not much later that Philip was thinking of a new navy of a hundred ships. By this time Attalos had gone home after wintering in Aigina. The joint Roman-Attalid attack on Aegean islands earlier in 208 had all been either failures or ineffective. The recapture of Tarentum decreased Rome's interest still further, and for two years little or no Roman activity in Greece took place.

This was the moment for the new, third, mediation attempt. It was clear that the two earlier attempts had failed largely because the Romans had pushed the Aitolians to be greedy or to be obdurate. Now that the Romans were absent the Aitolians were obviously feeling the pressure of possible defeat. This time there were at least five mediators: Ptolemy and Rhodes, as before, were joined by Byzantion, Chios, and Mitylene. It is possible that Athens was there also, but this is only a guess. The list is geographically very interesting. Ptolemy and Rhodes were probably still concerned that they should maintain their relative naval predominance in the Aegean, and the possibility of a new Macedonian fleet was quite sufficient to alarm them. Two of the others were located near that part of the Asian coast which was dominated by Attalos, and they came also within Ptolemy's sphere of influence, if they were not actually occupied by Ptolemaic garrisons. Byzantion was perhaps largely motivated by commercial considerations, but these were hardly absent from the concerns of the others, particularly the Rhodians. Above all, this was a very high-powered delegation. The inactivity of the Romans and Attalos emphasized the naval strength of the mediators as a group. There was, as in all mediations, a veiled threat that they might intervene on one side or another. The Romans might be able to ignore such a threat; the Aitolians, with or without Roman support, manifestly could not.

The Aitolians may have been ready to make peace, but neither the Romans nor, as it turned out, Philip were particularly keen. Then Philip made one of his lightning marches, through the mountains into the heart of Aitolia, where he sacked the federal sanctuary and Assembly meeting place at Thermon. No help came from Rome. One ally, Sparta, was soundly beaten by Philip's ally, the Akhaian League. The neutral mediators turned up again, and this time Aitolia and Philip were willing to agree a treaty. The terms were essentially a return to the position *ante bellum*. Sulpicius, finally awake to what was happening, protested, pointing to the clause in the Romano–Aitolian alliance which forbade either side to make peace without the other. He was, however, unable to promise any practical assistance (though by this time Rome was clearly able to spare forces, naval and military, had it wished to do so). The Aitolians made peace.[33]

The Roman war with Philip went on, but now, as at the start, only in Illyria. Rome, having failed to help Aitolia, now had to commit an army of 11,000 men and a squadron of thirty-five quinqueremes, under an ex-praetor, P. Sempronius Tuditanus, to the new campaign, which is a measure of the usefulness to Rome the Aitolian involvement had been.[34] Tuditanus was quite unable to recover the territory lost to Philip at the beginning of the war, and refused the chance to fight Philip's army. Another mediation, this time by a group of Epeirotes, produced a peace treaty which left Philip in possession of some of Illyria. Rome was thus free of the war in Greece, but was left with a grievance against both Philip and the Aitolians. (The Epeirote mediation had some force: if the Epeirotes had intervened they could have brought victory, at least in Illyria, to the side they favoured. No explicit threat was made, and the Epeirotes conducted a skilful diplomatic campaign, but the threat surely existed in the minds of everyone involved).[35]

The use of the Roman fleet in the Macedonian war had been wholly self-centred. Having persuaded first the Aitolians and then Elis, Sparta, and Attalos, to join in the war against Philip, Rome had quite cynically allowed them to do the main fighting, showing no interest in anything other than preoccupying Philip. This, of course, was par for the course in the Aegean – Ptolemy I had acted in this way during the great siege of Rhodes by Demetrios, and Ptolemy II had done the same in the Khremonidean War, when the victim was Athens. The fleet of 25 Roman quinqueremes had accomplished little beyond showing the flag and raiding here and there. It is no wonder that the Aitolians wearied of the war and made a separate peace.

On the other hand it has to be accepted that sending the Roman fleet to Greece was very effective in preoccupying its enemy Philip. It may well not have been his intention ever to invade Italy, but since he never had the chance that is only speculation. The presence of the Roman fleet in Greek waters made it quite certain he could not even try. For sending a fleet of 25 quinqueremes, only a tenth of its total fleet, Rome gained security on that flank of the war, even when Tarentum was under Hannibal's control, and even when a Carthaginian naval force could reach Greek waters. The Carthaginians seem to have been particularly unenterprising, in contrast to Philip, who made very good use of his small heterogeneous fleet; to Carthage also, as with Rome, the war in Greece, though it will have pained both Philip and the Greeks to realise it, was a minor sideshow.

Chapter 11

Roman Domination Established

In 207 an insurrection began in southern Egypt, in the region around the ancient city of Thebes. It was led by men who had been trained to arms by the Ptolemaic government in the great crisis of the invasion of Antiochos III ten years before. The peasantry of Egypt had finally paid enough taxes for such fripperies as the great ship whose uselessness was evident to all, and an appallingly extravagant court. Within a year they had organized their own government under a pharaoh and had an independent state.

This insurrection began the collapse of the Ptolemaic state, which was hastened by the accession of the child king Ptolemy V in 204, and by the internal palace disputes of the courtiers which followed. The surrounding states leaned in hungrily to strip off choice provinces for themselves. By that time Antiochos III was returning from his great eastern expedition; Philip V made peace with Aitolia soon after, and with Rome a year later, in 205. In the west the great Romano-Carthaginian War was coming to an end. All the Great Powers were becoming free and able to exploit the Egyptian collapse.

Philip was the first to do so. In 204 he returned to his project of building a new Macedonian navy. The old plan of a hundred-ship fleet had been started, but suspended, and now it was revived. He had partly solved the problem of finance by hiring pirates from Crete and elsewhere to raid vulnerable places for him. One of his hirelings, Dikaiarchos of Aitolia, raided in the Troad and the Kyklades, and coordinated matters with the Cretans.[1]

Philip was now concentrating on expanding his power in the Aegean area, where he had to deal with the three fleets of Rhodes, Ptolemy, and Attalos, all potentially hostile. The first was the most important. Philip sent an agent, Herakleides, to Rhodes city, where he set a fire which burnt part of the fleet in the dockyard;[2] the city was also thoroughly embroiled in a war with several of the Cretan cities which had sponsored Philip's pirates.[3] So Rhodes was neutralised so far as Philip was concerned. The Ptolemaic fleet was paralysed by the concentration of the home government on the rebellion inside Egypt, and probably by a lack of finance as well. In Samos in 201 all the Ptolemaic ships were laid up, by which time Ptolemy was at war and only lack of money could leave his armament unmanned in such an emergency.[4] Attalos of Pergamon had to devote much of his attention to his mighty neighbour Antiochos III, who was operating in Asia Minor during 204 and 203. Philip's main targets were the many islands and cities which were not under the control of any of these powers, and it was their preoccupations which gave him the opportunity to make his move.

The trigger for the combustion of this pile of tinder was the death of Ptolemy IV in 204. This was kept secret for some months, but when it was finally announced it caused an implosion of the government in Alexandria, with assassinations, plots, and conspiracies paralysing the whole structure.[5] Antiochos III, whose peace with Ptolemy IV expired on the latter's death, was now free to try again for Koile Syria whenever he wished, while in the Aegean the Ptolemaic outposts were now effectively defenceless and unsupported. Philip and Antiochos agreed to seize the Ptolemaic provinces (a very similar agreement to that which Philip had made with Hannibal), though Egypt itself was to be left alone.[6] Philip had by now evidently built up a fleet of some size and used it in a campaign in the Straits which brought him control of several of the cities in the area, including Kios, Perinthos, Chalkedon, and Lysimacheia. This gave him effective control of the whole Straits, except for Kyzikos and the vital city of Byzantion. It also brought him the immediate enmity of Rhodes, and in response to his success (and to his friendships with the Cretan pirates) Rhodes now declared war on him.[7] Antiochos, after a year's preparation, invaded Koile Syria in 202, conquering it systematically.

Philip's employment of pirates was not new, for such men had usually been available for hire in wartime, when they became called mercenaries. His innovation was to use them to collect loot for him in peacetime. His campaign into the Propontis also collected a number of cities, though he angered even more, including Byzantion and the Aitolians. The Aitolians appealed to Rome, pointing to the danger posed by Philip, and were greeted with complaints about their 'desertion' in the previous war, and a flat refusal to become involved.

To Philip this was a green light. With his new fleet he felt he now had a free hand against other powers in the Aegean area. His fleet is described by Polybios in the preliminaries to the later battle of Chios as consisting of 53 *cataphracts*, an unknown number of *aphracts*, and about 150 *lemboi*. In a description of the battle, there are listed a ten, a nine, a seven, and a six, so most of the rest of the *cataphracts* would seem to be quinqueremes; *trihemioliai* and quadriremes are mentioned also.[8] It is possible that some of these ships were survivors from the old fleet of Antigonos Gonatas, but it seems highly unlikely that any of these were still seaworthy after forty years out of use. It is best to assume these were the ships that had been started in 208 and completed by 203.

It was a big fleet, composed of a sensible variety of ships, but it was not enough. Philip raided and conquered in the Kyklades in 201, leaving garrisons on several of the islands, and came to Samos. This was a Ptolemaic island and naval base. Philip seized it and outfitted some of the Ptolemaic ships he found in the dockyard, taking them into his new fleet; how many ships he thus acquired is not known, but it was probably not many. His purpose was no doubt to increase his naval strength in the face of the enemies he had conjured up, whose aggregate joint fleets were greater than his. The main military attention in Egypt was now devoted to Antiochos' invasion of Koile Syria, and anything that could be spared from that conflict went to contain the southern rebellion. The Aegean ships had been laid up, probably because of a lack of money to pay the personnel; and the wars in Syria and Egypt were land wars, in which navies were not used. The Ptolemaic government apparently made no protest at Philip's actions: a protest might well have resulted in him annexing the island.

Philip's expedition had brought him fairly close to Rhodes, which already considered itself at war with him. A Rhodian fleet of perhaps twenty or twenty-five ships now came north, presumably aiming to deter Philip from approaching any closer to the home island. So when Philip nevertheless came south from Samos the two fleets met near the island of Lade. Outnumbered probably two to one, the Rhodians were quickly beaten, losing two quinqueremes captured and at least one seriously damaged. Most of their ships, however, got away.[9] Philip sailed into Miletos where he was welcomed, though the Milesians could scarcely do anything else.

This was the final straw for his many victims. The defeated Rhodian commander, Theophiliskos, contacted King Attalos and persuaded him of the danger of Philip getting still more territory. Attalos had been at war with King Prusias of Bithynia, who was Philip's ally and brother-in-law, and he was also apprehensive about the steady growth of the power of Antiochos III, Philip's colleague, though Antiochos III was busy in Syria for the time being. Chios, Byzantion, Kyzikos, and Kos now joined with the two major powers in a grand naval coalition. Before their fleets could assemble, Philip attempted to knock out Attalos, the most powerful of the coalition, by an invasion of his kingdom, but was unable to capture Pergamon. He retired and attacked Chios Island, thereby stationing himself between Rhodes and Kos to the south and Attalos, Byzantion, and Kyzikos to the north.

Philip's blockade of Chios city gave the allies time to gather their forces, and Philip tried to escape what had now become a trap by heading back to Samos. He was intercepted by Attalos' fleet, and a battle developed. The Rhodians, who had been further off, came up, and the battle grew into two linked fights, described by Polybios in a way which emphasizes the confusion of any battle. The Rhodians found their usual manoeuvring difficult because of the presence of the many *lemboi*; Attalos in his flagship became separated from most of his own fleet and was driven onshore; Philip's admiral Demokrates died early on; two of Attalos' commanders, Deinokrates and Dionysodoros, had to fight for their very lives. Philip allowed himself to be distracted into chasing after Attalos and plundering the riches of the grounded royal ship. The experience and skill of the allied fleets meanwhile gradually overcame Philip's men. When Philip returned it was to find that his fleet was being destroyed. He broke off the fight, gathered up his surviving ships, and next day reached safety at Samos.[10]

The two fleets had been approximately equal in strength, though Philip's many *lemboi* had given him a considerable numerical superiority. He also had more of the very biggest ships among his 53 *cataphracts*. The allies, however, had superiority in quinqueremes and the bigger ships, as well as greater nautical skill. Out of his fleet of 200 ships, Philip lost 24 of the *cataphracts*, including a ten, a nine, a seven, and a six; three *trihemioliai* were sunk and 65 of his *lemboi*; very few were captured. He therefore lost about half his fleet, mainly sunk. In this destructiveness the battle was unlike those in the west fought between Romans and Carthaginians. Here the aim was to sink, by ramming, which resulted in a lot of casualties by drowning; in the west the aim was to capture by boarding, at least in Roman tactics. The allies lost far fewer ships and men: four quinqueremes, two quadriremes, a *trihemiolia*, and a trireme were sunk, as was Attalos' royal ship, which was an eight. Even worse for Philip were the deaths of several thousands of his men, 9,000 killed and nearly 3,000 captured, according to Polybios (whose figures may well be exaggerated, but not by very much).

Philip's naval campaign should have been concluded at this point, but the allied fleets separated, leaving him free to move on. From Samos he followed in his stepfather's footsteps into Karia, where Olympichos had been his agent for a time, and where Philip now campaigned with a viciousness which had become habitual to him. The news of his agreement with Antiochos to partition the Ptolemaic empire became public, and provoked further alarm among the friends of Ptolemy, notably at Athens. The partition agreement suggested that Philip would go on raiding and conquering until he had subdued all the Ptolemaic posts, and many of the independent cities as well. He had, after all, enlarged his kingdom, expanded towards the Adriatic, subdued Aitolia, and gained control of the Straits and many of the Aegean islands. Now he was gaining territory in Karia, despite a great naval defeat. Rhodes reported the agreement to Rome, distorting the aim subtly so as to provoke the Senate. Attalos returned to the fight, also reporting and complaining to Rome, and his and the Rhodian fleets blockaded Philip over the winter in the small Karian town of Bargylia. He escaped in the spring and returned to Macedon with his remaining ships.[11]

From Macedon Philip sent an expedition against Athens, in which four Athenian warships were captured, though the Rhodians soon recovered them and sent them home.[12] Philip took his fleet on another raid into the Hellespont, where he captured the key port-city of Abydos after a long and nasty siege. With Lysimacheia this gave him control of both shores of the waterway – and on the way he had snapped up the Ptolemaic towns of Ainos and Maroneia in Thrace, the Thracian Chersonese, and the city of Sestos.[13] Rhodes meanwhile sent a squadron to take over many of Philip's Kyklades conquests, and sent other ships to help at Abydos, where troops and ships from Attalos and the nearby city of Kyzikos were also helping.[14] (Kyzikos, apart from Byzantion, was almost the only city in the Straits not under Philip's control; no doubt it realized it was probably next on his list.)

Rome finally decided that Philip was a danger, and at Abydos he was presented with an ultimatum. He ignored it, and was therefore all the more surprised to find that not only were the Romans serious, but that they had now placed themselves at the head of a coalition of the many Greek cities and states he had insulted, angered, or damaged in the past five years.

Furthermore, Rome had a fleet which put his in the shade. Two legions with auxiliaries, over 20,000 men, were carried across the Adriatic with fifty quinqueremes, twenty of which were at once sent on into the Aegean to help Athens.[15] Philip's mode of warfare was now mainly raiding, and Athens was subject to a long series of raids mounted from Chalkis in Euboia and from Corinth; the arrival of the Roman squadron, when allied to local Athenian ships and those contributed by Attalos and Rhodes, provided enough power to mount a reprisal raid. A surprise attack captured Chalkis and destroyed Philip's supplies there.[16] This increase in Athenian resistance brought Philip himself south, but by land. The allied squadron had now secured command of the southern Aegean as well as the eastern part of the sea, which had been the result of Philip's return to Macedon from Karia. The Rhodian ships in the Kyklades, by removing Macedonian authority from most of the islands and establishing that of Rhodes instead, pushed Philip's maritime boundary further north. Philip continued to hold Andros, Kythnos, and Paros, as well as the whole coast from the Malian Gulf to the Propontis. Further, Philip found that he had almost no allies at all in Greece; even his old friend the Akhaian League decided to concentrate on fighting Nabis of Sparta rather than helping him; his only active ally were the Akarnanians, who had been the cause of his quarrel with Athens in the first place.

The Romans' strategy was to base the army in Epeiros and try to force a way into mainland Macedon through the mountains. Their naval base for this was once more Corcyra, but to leave the ships there while the army campaigned in the hills on the Macedonian border was to have them do nothing. Next year (199) the whole fleet was sent to the Aegean, under the command of the legate L. Apustius. He took it on a cruise to Andros, which was captured and given to Attalos, to Kythnos, which successfully resisted the Roman attack, then north to raid Skiathos and menace Kassandreia and the Chalkidike peninsula, part of the Macedonian homeland.[17] None of this would seriously affect Philip, but at the end of its cruise the fleet fetched up at Oreus, on the north coast of Euboia, to which it laid siege.

This last move had useful effects. It was one of the factors which helped persuade the Aitolians to join in the war on the Roman side, for the Romans now found they needed their help. The fleet at Oreus included Rhodian and Attalid ships, and a few Athenians. While the siege was on, troops from the fleet conducted raids on the nearby mainland. The places which were captured were soon lost again, but Philip was annoyed, and Oreus fell.[18] It was retained, being yet another of Attalos' acquisitions. Oreus in allied hands helped to block further Macedonian raids southwards, as did control of Andros and the Kyklades.

While all this was going on the Macedonian fleet lay at its main base at Demetrias, only a day's voyage north of Oreus, commanded by Herakleides, the firebug of Rhodes. He had also held a high command under Philip at the battles of Lade and Chios. The two rival fleets were about equal in numbers: Herakleides had twenty-five *cataphracts* and eighty *lemboi*;[19] Apustius had seventy warships and twenty smaller vessels, but the disparity in strength as opposed to numbers was too great for the Macedonian fleet to risk coming out. Apustius' quinqueremes were decisive. Philip's fleet may also have been undermanned, for he needed all the troops he could collect to defend his frontiers.

The next year – the allied fleet had dispersed for the winter – the three squadrons rendezvoused at Andros. The Rhodians sent twenty ships, Attalos sent twenty-five, and the Romans probably had as many as these two put together. Command was with L. Quinctius Flamininus, the brother of Titus, the consul of 198, who commanded the armies in the west. The joint fleet campaigned once again in Euboia, taking Eretria and Karystos to add to Oreus, though Chalkis remained under Macedonian control. From there Flamininus then sailed to Kenchreai, the eastern outpost of Corinth.[20]

This move was as much political as naval. The city of Corinth and its acropolis Akrocorinth were held by a Macedonian garrison, but this was a place which the Akhaian League most wished to recover. The prospect of it falling into Roman hands, or that of Attalos (who had already acquired Aigina, Andros, and Oreus as part of his campaign agreement with Rome) was something which might help persuade the Akhaians to join in the war. There was a three-day meeting of the Akhaian council on the question, which eventually, with some reluctance, decided to ally with Attalos and Rhodes, and to pursue an alliance with Rome. Meanwhile Flamininus had actually captured Kenchreai. All then joined in a siege of Corinth, but the resistance was determined, both by the citizens and the Macedonians, and winter was approaching. The fleets went off to their winter quarters again.

The fleets were not used next year, for by defeating Philip as early as June, as T. Flamininus did at Kynoskephalai, it became clear that no naval campaign was needed. And those allies in Asia Minor and the Aegean islands were now inevitably distracted by events in the east. While this war in Greece was continuing, Antiochos III had finished off Ptolemaic resistance in his campaign in Syria. The final part of the conflict had been a siege of Sidon, where the last Ptolemaic army, commanded by Skopas of Aitolia, held out for a time. But there was no hope of relief, and in 198 Skopas surrendered. Antiochos turned away from Egypt, returning to North Syria.

It may be that Antiochos had captured some Ptolemaic ships in the several Syrian ports he had now taken – Tripolis, Byblos, Tyre, Sidon, Ptolemais-Ake, and so on – though there is no indication of this in the rather poor sources for this war. By the beginning of the campaigning season of 197, however, he had gathered a substantial fleet of a hundred warships and 200 *lemboi*.[21] His army was sent to march overland into western Asia Minor, and he himself took the fleet westwards along the coast of southern Asia Minor.

The size of the fleet suggests that an intensive building programme had been under way for the last year or two. Antiochos had captured some ships from Ptolemy IV in the Fourth Syrian War in 218. They had thus been built before then, and by 197 were well over 20 years old. They could be just about usable still, but until now there had been no mention of any other fleet under any Seleukid king, and no mention at all of *lemboi*. This type had been introduced into Greek shipbuilding terminology by Philip in the last few years; Antiochos had clearly accepted their usefulness. We must assume that the great majority of the ships were newly built, during the years since Antiochos determined on his new war in Syria – that is, since 203 – though it is also possible that some had been built earlier.

His purpose in campaigning along the coast from Syria westwards was to mop up the cities and posts still held by Ptolemy in Asia Minor, all of which were on or close to the coast. The campaign was almost entirely unresisted, involving only one siege, lasting a month, at Korakesion. Antiochos ignored the rest of Pamphylia, where the cities were not under Ptolemaic control, and sailed right on across the bay to a small anchoring place at Korykos, just north of Cape Gelidonia. The Rhodians sent an embassy to find out his intentions, but Antiochos first reached the island's neighbourhood quicker than the Rhodians expected, and then pacified them by agreeing that it should exercise a sphere of influence over the Karian coastal cities and the islands as far north as Samos.[22] Rhodes now had a nice little empire, consisting of some of the Kyklades, Samos, and several of the Karian cities; more would be added in the peace settlement being made with Philip in Greece.

Antiochos arrived at Ephesos with his fleet.[23] At once his hundred warships made him the dominant naval power in the Aegean. Rhodes' fleet was usually less than fifty ships; Attalos had never put more than thirty-five to sea at any one time; Philip had less than twenty-five left; and the Roman fleet was now based once more at Corcyra. Antiochos' army had arrived at Sardis, and it had encroached somewhere on the territory which had been occupied by Attalos of Pergamon. Attalos himself died about this time, and was succeeded by his son Eumenes II. There was therefore some hesitancy in Pergamene policy, just at the time everyone in Greece was preoccupied with the aftermath of the Macedonian war and with Antiochos' arrival. Antiochos was in contact with Rome and with Flamininus, whose victory at Kynoskephalai had come not long before Antiochos' arrival at Ephesos. Between them Flamininus and Antiochos established a new political system in the Aegean, effectively eclipsing all other powers.

Ignoring what was happening in Greece, Antiochos and his fleet went on northwards. He took over those places in the Hellespont and the Propontis which Philip had seized from Ptolemy a few years before. Philip had apparently withdrawn his garrisons and no one had filled the vacuum. Lysimacheia had been abandoned and was in ruins, its population scattered or enslaved. This was the work of the Thracians, who had presumably captured the city between Philip's abandonment of it and Antiochos' arrival. Antiochos campaigned several times in the next years to establish his control in Thrace, and he restored Lysimacheia, rebuilding the city and searching out and ransoming the people. In the process he became the lord of the Hellespont and the Straits.[24]

During 196 his representatives in Syria negotiated a peace treaty with the government of Ptolemy V. This formalized the transfer of Syria and the Asia Minor and Thracian territories which had formerly been held by Ptolemy to Antiochos' rule. Ptolemy kept Egypt, Cyrenaica, and Cyprus, but of his other overseas lands he held on to Itanos, Thera, and Methana alone, though this does suggest that he continued to maintain a naval presence in the Aegean, as he clearly did in Egypt and Cyprus.

The reorganization of Greece to Roman satisfaction took several years. Apart from negotiating the transfer of lands from one state to another, it turned out to be necessary, at least in Roman eyes, to campaign against, and reduce the power of, Nabis of Sparta. This required the attention of the fleet from Corcyra, which assaulted Sparta from the sea, while the proconsul Flamininus and the Akhaians attacked it from the landward side. But then the Roman army and the fleet returned to Italy.[25]

Negotiations between Rome and Antiochos had continued from the year 200 to 193, with envoys passing back and forth. Gradually it became clear to both sides that their differences were irreconcilable. The essential problem, though scarcely voiced, was that both wished to be supreme, and neither could accept the other as an equal. Antiochos had established his power over much of Asia Minor, but in the process he had clashed with both Rhodes and Eumenes of Pergamon. In Greece, Rome's settlement had left several states discontented at the details of the settlement, notably the Aitolians. These several quarrels and resentments coalesced.

In 192 Antiochos was invited into Greece by the Aitolians, with the intention of overturning the Roman system, recently established. He perhaps assumed that Rome's withdrawal was definitive and that he could go to Greece and establish his own hegemony without Rome reacting. He travelled to Greece with an army which was scarcely more than a nominal force: 10,500 soldiers, carried in a fleet of 200 transport ships escorted by 40 warships and 60 lighter vessels – about half of his war fleet.[26] Within a few weeks he had brought much of central Greece into his protection; to the north was Philip, still ruling in Macedon, whom he must have hoped would join him; to the south was Athens and the Akhaian League, neither with much military power, but now with memories of good Roman support in the recent past. None of Antiochos' wars had been so easy, and if that had been all that was needed, he had judged to a nicety what force he required. But this was not all.

Rome was scarcely going to allow her clients in Greece to be filched in this way. In addition, there had been at Antiochos' court for several years the one man Romans feared most: Hannibal, driven into exile from Carthage, had fled to Phoenicia and was received by the king. Stories of his supposed influence on Antiochos, which in fact was no more than any other courtier, were eagerly believed in Rome. When Antiochos took over central Greece and was known to have an army and a considerable fleet with him, Roman imaginations conjured up a plan by which he would land 10,000 men, 1000 cavalry, and 100 ships in Italy, and assumed that this was what Hannibal was advocating. Such stories may actually have been put about simply to frighten the population into supporting yet another war; they were certainly not serious plans – unless the Romans were still nervous about the loyalty of the Greek cities of the south, and of the Samnites and others in the Italian interior, many of whom had joined Hannibal earlier. The very numbers indicate that it was not a serious plan but a Roman fairy story.

Rome made serious preparations, partly defensive, but mainly aggressive, in case an invasion of the west was really intended. This was, after all, the first time since the Illyrian Wars that a major enemy had controlled a substantial part of the western Greek coast. There were numerous harbours and even major cities which could be the launching points for an invasion force, or at least sally ports for raiding squadrons. The magistrates for 191 were assigned to the provinces and cities facing the danger: Sicily, Bruttium, Tarentum, and Brundisium – and Sicily was to have two commands, one for the south coast and one for the east. The fleet was mobilized, orders were given to build twenty more quinqueremes and enlist crews from the naval allies, twenty ships being assigned to defend Sicily, and fifty to Greek waters.[27]

Rome evidently apprehended possible trouble from Carthage, judging by the Sicilian dispositions, but envoys soon arrived from that city, and from King Masinissa of Numidia, from Ptolemy V, and from Philip V, all promising gifts or forces or co-operation. Meanwhile the Roman fleet in the Ionian Sea supervised the transfer of the consular army to Greece, and then ranged south along the coast, where its very presence encouraged resistance to Antiochos and the Aitolians at several points.[28]

The key member of this wide anti-Antiochos alliance was Philip. He allowed the Roman army to march through Macedon, supplied its needs, and sent troops to assist it, hoping to recover some of the lands he had lost as a result of the previous war. Antiochos withdrew his advanced posts, and found that the Aitolians were of little help. He was defeated at Thermopylai and then evacuated his troops by sea. By this time part of the Roman fleet had come round to the Aegean and ambushed part of a convoy of supplies near Andros, but did not attempt to interfere with the evacuation.[29]

With Antiochos back in Asia, and the Romans victorious in Greece, the war became a naval contest. The consul M. Acilius Glabrio, the victor at Thermopylai, turned back to deal with the Aitolians. The praetor A. Atilius Serranus, in command of the Roman squadron, went into Peiraios. Glabrio could not get across the Aegean, nor could Atilius venture very far, for he had only twenty-five ships. So Antiochos had several months in which to prepare for the Roman invasion of Asia. That he assumed this would take place is shown by his early visit to Lysimacheia at the Hellespont, no doubt to check on preparations and perhaps to install reinforcements. He sent Hannibal to Syria to organize a new fleet, which was to sail to the Aegean next year.

Atilius' successor as naval praetor for 191/190, C. Livius Salinator, was assigned more ships, but he had problems collecting and manning his force. He started out from Ostia in the spring of 191/190 with fifty *tectae* and twenty-four *apertae* which were picked up at Neapolis. These are Livy's terms; Polybios uses *cataphract* and *aphract*, and probably refers to quinqueremes and either triremes or *lemboi*, probably the former. The ships of the naval allies had been collected from several cities, which had also supplied the crews. More of these ships joined at Rhegion and Lokroi, as did six Carthaginian ships.[30] All this took time, so Livius did not get to Aegean waters until about June, though he did arrive with fifty-six quinqueremes and about thirty smaller vessels. He was met both by Atilius and his twenty-five ships and by King Eumenes with three ships, probably quinqueremes, giving him a fleet total of eighty-one quinqueremes and about thirty smaller ships.[31]

Eumenes had spent the winter of 192/191 at his island of Aigina. Some of his soldiers had been involved on the Roman side in the land fighting in Greece, but he had carefully avoided an open quarrel with Antiochos, for his kingdom was very vulnerable to the Seleukid army in Asia Minor, especially now that

Antiochos was back at Ephesos. The three ships he had with him were little more than a royal guard, and the rest of his fleet was presumably laid up at Elaia, his main port. By publicly greeting Livius, however, he had now aligned himself openly with the Romans.

On his way from Chalkis to Ephesos with his fleet and defeated army, Antiochos had sailed past Andros, Eumenes' island, so preserving the fiction that they were not fighting each other. But he had stopped at Tenos, the next island after Andros. This was the site of the headquarters of the Island League, which was now under Rhodian hegemony. Rhodes and Antiochos had had friendly relations in 197, when he sailed west, and Rhodes had remained neutral in the Roman war so far. By calling at Tenos, Antiochos was making a political investigation, trying to find out Rhodian intentions. There was some fighting between Seleukid and Nesiotic troops, but this did not necessarily imply Rhodian hostility, any more than the participation of Attalid troops on the Roman side in the mainland war had done. Both Eumenes and Rhodes were treading very carefully to avoid an open conflict with Antiochos until credible Roman forces arrived. And Antiochos was just as carefully avoiding a conflict, possibly hoping the others would remain neutral.

The arrival of the Roman main fleet made Rhodes's participation all the more valuable, since that city now held the naval balance: whichever side Rhodes joined would acquire the cooperation of another forty or fifty ships and of the highly skilled Rhodian sailors and commanders. The possibility was no doubt one of the reasons Antiochos had ordered up another fleet from Syria, to be collected, built, and commanded by Hannibal.

Rhodes made its decision between Antiochos' retreat and Livius' arrival, and had publicly joined Livius by August. He arrived at Peiraios in June, spent two months preparing, organizing and negotiating, and moved to Delos in August. (This is the obvious sign that Rhodes had joined him, for Delos was part of the Island League; no public announcement seems to have been made of Rhodian belligerency.) When he had returned to Asia, Antiochos had at once taken his fleet to the Hellespont – which at the time was the obvious danger point – but when warned that Livius was at Delos he returned to Ephesos.[32] Meanwhile, King Eumenes organized the launching of his ships at Elaia, twenty-four quinqueremes and twenty-six smaller vessels.[33] Antiochos' admiral Polyxenidas had 70 quinqueremes and 130 smaller ships in his fleet.[34] This is a smaller number than Antiochos had come west with; some may have been lost in the meantime, but the main reason for the reduction is probably that a squadron was left at the Hellespont.

Antiochos thus faced a seaborne attack from across the Aegean, seemingly aimed directly at his political centre at Ephesos, and a likely land attack by way of Macedon and Thrace aimed at crossing into Asia at the Hellespont. The naval war, however, was principally aimed at securing the Hellespont for the army, not Ephesos, though a threat directed at that city certainly gained

Seleukid attention. Two other Seleukid enemies were close by: Eumenes' main fleet was coming out of Elaia, and a Rhodian fleet was coming north. The land attack could not happen while the Aitolians were preoccupying the consular army in Greece, and then could not reach Asia unless the fleets had secured the Hellespont crossing. So it was the naval attack which had to be countered. And, since the three enemy fleets were separated, it was necessary to keep them apart.

Polyxenidas stationed himself first at Phokaia, between Eumenes and Livius. Livius brought his fleet across from Delos to an anchorage in southern Chios, thereby threatening Ephesos directly. Polyxenidas came south to station himself at Kissos, on the south coast of the Erythaian peninsula, between Livius and Ephesos, defending the approach to the city. This allowed Eumenes to follow him and to join Livius. The allied fleet came cautiously eastwards and was confronted by Polyxenidas not far from his base. Livy has a clear description of the preparations by the Romans for the fight, hauling down the sails and ordering the formation, and of the confusion involved. (He normally describes the change from one order of sailing to another as 'confusion', which is not necessarily how the sailors would have described it.) Polyxenidas concealed the size of his force for a time, and sent three fast ships to attack two Carthaginian quinqueremes which Livius had sent out in advance – one of these was taken, but when the Seleukid ships attacked Livius' own ship they were seized by grappling irons (apparently a more flexible version of the discarded *corvus*) and boarded. In the action which followed the Romans attacked the Seleukid left wing (the sea wing) first, and Eumenes came along to attack the right, close to the coast, a little later. The heavier Roman ships now had their effect, as did the allies' greater numbers. Polyxenidas broke off the fight and got most of his ships away, being lighter and faster. But he had lost ten ships swamped (presumably rammed and holed) and thirteen captured.[35]

Polyxenidas got back to Ephesos first. Livius and Eumenes were now joined by twenty-four Rhodian ships which had been waiting at Samos. The joint fleet, now even stronger than ever as against the reduced Seleukid fleet, demonstrated outside Ephesos harbour, but could not attack in the face of Polyxenidas' fleet supported by the city garrison and the Seleukid forces. It was by now September; everyone went to winter quarters, the Rhodians at home, Eumenes at Elaia, and Livius at Kanai, just north of Elaia.[36]

The result of the Battle of Kissos (or Korykos) was scarcely decisive, particularly in view of the naval building going on through the winter in Syria. (Livy places Hannibal's appointment to this task after Kissos, but it seems much more likely it had taken place earlier; the arrival of Livius' force, together with the clear enmity of Rhodes and Eumenes, which made it clear that Antiochos' fleet was outnumbered, would be the most likely time). Polyxenidas was also set to work building and equipping more ships at Ephesos. Antiochos himself spent the winter inland, recruiting allies and new forces for the army.

It was clear that a further naval campaign was going to be necessary. Whether news from Rome had arrived is not known, but it is very likely – Rome certainly received news of Antiochos' own naval build-up, so it is probable that news went the other way as well. The Romans had made L. Cornelius Scipio the consul to command the invasion of Asia, and he was to be accompanied by his formidable brother Africanus. They were to have a new army, distinct from that in Greece, but could collect whatever troops could be spared from Glabrio's army, which was still fighting the Aitolians. The fleet was to be commanded by L. Aemilius Regillus as praetor, and he would take with them a reinforcement of twenty ships, 1,000 sailors from the naval allies, and 2,000 infantry. When the Senate heard that Antiochos was building ships it ordered the construction of another thirty quinqueremes and twenty triremes at Rome.[37]

It took several months for the Scipios to get their army to the Hellespont. The Aitolian war had to be wound up first, for they needed Glabrio's army as well as that they had brought from Italy. In addition they surely wanted the naval conflict to be resolved first. If Antiochos could win in the sea war he could probably prevent an invasion of Asia by blocking the Hellespont crossing, or by reviving the war in Greece; if he lost at sea he would need to fight a land battle against an army and a general which had already beaten Hannibal, Philip, and the Aitolians. Clearly this encounter was something the king would wish to avoid if possible.

In the spring, Livius took thirty of his and four of Eumenes' ships to the Hellespont to try to clear the way for the army.[38] The rest of the ships were left at Kanai under the command of the Rhodian Pausistratos. They were positioned so as to prevent the Seleukid fleet at Ephesos from reaching the same destination, but at the same time it left the smaller islands open to Seleukid attack. Rhodian posts were thereby vulnerable. Sure enough Pausistratos was decoyed away to Samos by Polyxenidas, who then raided the Rhodian squadron at Panhormos on the north coast of the island, capturing twenty of the ships and sinking one. Seven ships survived and escaped southwards.[39] At about the same time Antiochos' son, Seleukos, captured Phokaia, which had been used as a base the year before by the Romans. When the news reached Livius he broke off the Hellespont campaign and returned to Kanai. He and Eumenes came out, joined their squadrons and moved to Phokaia. The Rhodians, annoyed at being caught napping at Panhormos, appointed a new admiral, Eudamos, and sent out two successive squadrons as they became ready, ten ships in each, to bring the surviving ships at Samos up to their former strength.[40]

Polyxenidas in Ephesos was now again situated between the Roman-Attalid fleet and the Rhodian squadron, but he was still outnumbered by Livius' force. Livius brought his fleet south round the Erythraian peninsula and prepared to cross the open sea from the mainland to Samos to join the Rhodian ships. This was a voyage of about 40 km. Polyxenidas came out to try to intercept, waiting at Myonessos, off to the east. The allied fleet was scattered in its crossing by high seas, but Polyxenidas was also prevented by the same storm from getting to grips with them. Some of the allied ships did reach Samos, but discovered that Polyxenidas was waiting to the south of the island to make another interception attempt. They returned to the mainland, and warned their commanders. Polyxenidas, whose ships had been at sea by now for several days without rest or the chance of refreshment, had to return to Ephesos. This gave the allies the opportunity to cross to Samos, where they were joined by the new Rhodian squadron soon after.[41]

Antiochos' fleet naturally refused to come out of Ephesos when the allied fleet, which now greatly outnumbered it, paraded before the harbour. The new praetor, Aemilius, now arrived to take command. His voyage had been uneasy. Antiochos had contacted a number of 'pirates', so-called by the Romans, who conducted raids. One group operated out of Kephallenia off the west coast, where they might cut sea communications with Italy.[42] Others operated out of the Hellespontine cities, raiding as far as the Kyklades.[43] The only evidence of their piracy is that they attacked Roman and allied ships and lands, so they were probably in actual fact subjects or allies of Antiochos. In crossing the Aegean, Aemilius' two quinqueremes were escorted by two Rhodian and two Italian triremes, and met two Rhodian quadriremes, two Roman quadriremes and two Attalid quinqueremes on the way. To need a force of four quinqueremes, four quadriremes, and four triremes in such a voyage does imply that something more than a few 'pirate' ships were loose in the Aegean. Since Aemilius is said to have set forth from Italy with twenty ships, it would seem that he had left most of them behind to secure his communications.[44]

A conference of the allies at Samos revealed contrasting priorities: Eumenes' lands had been invaded and he was naturally concerned to protect them; the Rhodians were concerned at a Seleukid force operating against their lands on the mainland, and were keen to get control of Patara, which had a large harbour ideal for the accumulation of a fleet intending to invade their island. And it was known that Hannibal was about ready to move westwards with the new Seleukid fleet, directly threatening Rhodes itself.[45]

This last was the crucial matter. Hannibal's precise instructions were not known, but it could be that he intended to attack Rhodes. Even if he did not, the arrival of his fleet would be a serious reinforcement for Antiochos in the Aegean. So it was necessary to block his progress before he reached Rhodian waters. A large part of the Rhodian squadron at Samos returned home, collected such ships as were still there and went east to join another squadron which had been sent on already. They all joined at Megiste Island, just west of Cape Gelidonia, and sailed round to Phaselis. From there they moved on to the Eurymedon River, where they learned that Hannibal's fleet was at Side, only a few kilometres further on.[46]

Hannibal had a fleet of almost fifty ships, thirty quinqueremes, and ten smaller ships, perhaps triremes, and three sixes and four sevens.[47] This is a different sort of fleet than Antiochos had brought west on his original voyage. The sixes and sevens had presumably been built over the winter at Hannibal's instructions, perhaps with the experience of the bigger and heavier Roman quinqueremes at Kissos in mind. The Rhodians had thirty-two quadriremes and three triremes in their force. Given the smaller and more manoeuvrable Rhodian ships and the more powerful Seleukid vessels, the contest was roughly equal. As the Rhodians sailed eastwards they met the Seleukid fleet arrayed in battle order, formed themselves up in response, though this is said to have resulted in some disorder in the Rhodian ships – Livy's 'confusion' again.

In the battle both sides suffered, and the Rhodians captured one of the sevens. Hannibal withdrew with only twenty of his ships undamaged, according to the Rhodian account Livy was using. The Rhodians followed him for a safe distance, thus allowing them to claim a victory. (Rhodian casualties are not stated.) And indeed they had successfully prevented Hannibal's fleet from reaching the Aegean. They did not, however, stop him from reinforcing Antiochos by land. The Rhodians withdrew all the way back to Rhodes,[48] leaving Hannibal free to operate in the Bay of Pamphylia: from there he was able to send troops north to Antiochos at Sardis. Antiochos would have preferred the ships, no doubt, but Hannibal's fleet was still a threat, and the Rhodians had to set a guard of twenty ships at Megiste Island just in case. As a result they could only send seven ships north to join their allies.[49] (This total of only twenty-seven ships seems to imply that the battle of Side had left eight Rhodian ships out of action.)

At Samos, Aemilius' fleet ran out of supplies, which were stockpiled at Chios. Antiochos had laid a trap, attacking the small town of Notion in the hope that the Roman-Rhodian fleet would come across to rescue it; the Seleukid fleet under Polyxenidas was also taking part. Aemilius heard of a consignment of wine at Teos which was intended for Antiochos and decided to seize it before going on to Chios. Polyxenidas learned of the move and took the Seleukid fleet to Megiste Island – not the one the Rhodians were guarding, but another which was about 15 km from Chios, close to the Myonessos Cape. Word reached Aemilius of the royal fleet and he brought his own fleet out of harbour and rowed to meet it.

The disparity in numbers was slightly in favour of the Seleukid fleet for once. The Romans had fifty-eight and the Rhodians twenty-two ships in the line (that is, quinqueremes and quadriremes); Polyxenidas had ninety, of which three were sixes and two sevens. These last are heard of here for the first time, and so they had been built during the winter at Ephesos, as another part of the naval build-up, along with Hannibal's fleet. Given that Polyxenidas' fleet had been reduced from a hundred to seventy-seven at Kissos, to which twenty Rhodian captures had been added at Panhormos, very little other new building had been done. The absence of Eumenes' fleet, and of some of the Roman ships from the allied force, was due to the fact that he had gone to the Hellespont to resume securing control there for the army to pass. Polyxenidas would thus have benefited very much from the presence of Hannibal's fleet. It would have given him overwhelming numbers; as it was, he had approximate equality only.

The allies had another weapon, a fire-pot, mounted on some of the ships on a pole in front of the bow, where it could be dropped on enemy ships. The very sight of it frightened everyone on board the wooden, inflammable, galleys. It had allowed several of the Rhodian ships to escape at Panhormos, and at Myonessos ships carrying the weapon were avoided by the Seleukid ships. It was, however, the formation adopted by the allies which was decisive in the battle: a compact force of ships was aimed at the centre of the Seleukid line, which was thinner because it was organized in an attempt to outflank the allied line. The heavy attack at the centre broke through, and the ships could then turn to take enemy ships in the rear in a classic *deikplous* manoeuvre.

The Seleukid fleet lost forty-two ships, thirty-two of them sunk or burnt – roughly half of the fleet. The allies lost only three. The Seleukid survivors retired to Ephesos, outside which Aemilius again paraded his fleet to emphasize his victory.[50] Both sides recognized the decisiveness of the defeat. Aemilius was now able to send thirty of his ships to the Hellespont, keeping the rest, some forty ships, to watch Polyxenidas' ships, and the Rhodians chose to go there also in order to emphasize to the consul their contributions to the victory (which are somewhat exaggerated, even invented, in the versions reaching the historians). Aemilius then took the rest of the fleet to Phokaia, which he recaptured and where he took up winter quarters.[51] He was thus in a position to prevent the surviving Seleukid fleet from sailing north to interfere with the army passing the Hellespont; the Rhodians were probably not strong enough to block any southward move, but for the present that did not matter: if Polyxenidas did come out and sail for Syria, the Seleukid fleet would be out of the way. Antiochos recognized the situation by withdrawing his troops from Lysimacheia and refusing to dispute the crossing.[52]

The Roman army therefore crossed into Asia and met and defeated Antiochos' army in battle at Magnesia. When he heard the result of the battle Polyxenidas tried to save the fleet, coming out of Ephesos and sailing for Syria. He was unable to pass the Rhodians stationed at the southern Megiste Island and took refuge in Patara. Unable to get out, and probably unable to feed his men, they all set out to get home to Syria overland; unfortunately we know nothing else of what must have been a difficult journey, to put it no stronger.[53] In the peace treaty which followed, Antiochos' fleet was surrendered; he was permitted to keep only ten ships, and these were not to be sailed further west than Kilikia. The ships at Patara – fifty of the quinquereme size and bigger –

were burnt.[54] Eumenes was given most of inland Asia Minor, and Rhodes gained the Karian and Lykian lands adjacent to their island.

Once again Rome had defeated a rival Great Power, the third in fifteen years. As with the final campaign in Africa and the defeat of King Philip, this had been achieved by the application of only a part of Rome's military and naval resources. Particularly impressive had been the intelligent strategic use of the Roman squadron by Livius, whose positioning of the ships at the harbour in southern Chios had given him control of the central position, and from this all else flowed: the victory at Kissos, the ability to reach and eventually control the Hellespont, and the decisive separation of the two Seleukid fleets. The obvious move for him would have been to sail to the Hellespont, but this would have left him vulnerable, short of supplies, and bottled up, probably in the Propontis, while almost certainly both the Attalid and the Rhodian fleets would have been overwhelmed by the bigger Seleukid fleet.

The battles are interesting also in that only in the last, at Myonessos, were there really heavy casualties. It was only when Antiochos' fleet had been decisively outnumbered, by losing so many ships at Myonessos, that the sea war came to an end. No fleet was ever going to fight an enemy fleet which was its decisive numerical superior; battles only took place between approximate equals, who each thought they could win. This, of course, was why Philip did not use his fleet against Rome, and why he fled from even the rumour of the approach of Roman quinqueremes: defeat in battle was certain in a fight between quinqueremes and *lemboi*.

When it came to battle the result arose from the initial formation, from skills in manoeuvring, and from the weight of the ships. None of these was enough in itself. At Side, for all the Rhodian skills and later boasts, the battle was effectively a draw, though it was also a strategic victory for the Rhodians; Kissos was a Roman-Attalid victory that could not be followed up to make it decisive because the Seleukid ships were quicker and could get away to safety. The right combination came at last at Myonessos: the right battle formation bringing a breakthrough in the centre (the opposite to what had been attempted at both Kissos and Side), and the skilful use of manoeuvre (not just by the Rhodians) – all these together neutralized the larger and more numerous Seleukid ships.

Chapter 12

Roman Domination in Action

In 221, when the Fifth Syrian war began, there were four major naval powers in the Mediterranean: Egypt, Rome, Carthage, and Macedon, each with a seaworthy fleet of a size capable of fighting a major war; less important naval powers were Rhodes, the Attalid kingdom, and the Seleukid kingdom, with fewer ships, but able to dominate their local seas. By 188, when Roman commissioners dictated peace to Antiochos III at Apamea in central Asia Minor, Rome was the only naval power left, with Egypt, Rhodes, and the Attalid kingdom still having minor local roles. This had happened within a single lifetime, within the reign of Antiochos III (223–187).

This decisive change in naval power had been partly accomplished by Antiochos III by his destruction and appropriation of the Ptolemaic overseas possessions in Syria and Asia Minor. This had not, however, wholly destroyed the Ptolemaic navy, though it had severely reduced its range; much of the damage had been done by the wrongheaded and negligent Ptolemaic policy in the past two or three decades (such as concentrating resources on the useless gigantic forty).

In the rest of the Mediterranean the alteration was mainly the work of Rome. That city's successive victories over Carthage, Philip V, and Antiochos III were each concluded by a peace treaty in which the defeated state was obliged to abandon any pretensions to naval power. Carthage had to surrender all her ships except ten triremes; those which were surrendered were publicly burned by Scipio Africanus, accompanied by the lamentations of the Carthaginians.[1] Philip V was allowed to keep a huge sixteen, for sentimental reasons – it had probably been one of Antigonos Gonatas' ships, and maybe had originally belonged to Demetrios Poliorketes; otherwise Philip was allowed to keep only five small ships, probably triremes.[2] Antiochos III was obliged to maintain no more than ten triremes and ten smaller ships, which were not to be sailed west of the western boundary of Kilikia, and the fifty ships Polyxenidas had taken from Ephesos and had tried to save were burned at Patara.[3]

Other than the two cases of the burnt ships, we have no information as to whether these terms were really carried out. Antiochos III, for example, had fleets in two places, that under Polyxenidas in Ephesos which was partly sunk at Myonessos and the rest burned at Patara, and the fleet which had been gathered and built by Hannibal in Syria and which had fought at Side; if Polyxenidas at Ephesos was trying to save his ships by sailing to Syria, it is highly likely that the ships off Pamphylia were also moved home. We know of the fate of the Patara ships, but nothing is known of the fate of the Syrian ships. They were supposed to be surrendered to the Romans, but this is never stated to have happened, nor is any Roman recorded as going to Syria to see to the ships' destruction, though two men did go to Syria to take the oath of the king to observe the peace terms. (It does seem very likely that Livy would have mentioned such an event if it had occurred.) For all we know the ships were simply kept by Antiochos, perhaps dispersed among various ports. It is not difficult to imagine the other defeated states actively avoiding compliance with the Romans' terms as well, though it may have been more difficult for Carthage, only a day's sail from Sicily.

Further, after complying with, or evading, the terms of peace at the time, nothing seems to have stopped Rome's defeated foes from building new ships. Already in 192 Carthage was able, without a blush, to offer six quinqueremes to assist Rome against Antiochos, and, without a comment or a complaint, Rome accepted them.[4] This was from a city which only nine years earlier had been instructed to have no warships bigger than triremes. And if Carthage could do this then there is no reason to suppose that other powers were any more compliant.

Not that, in naval terms, this mattered very much. The experience of being beaten by the Romans was sufficiently unpleasant for most states not to wish to provoke another bout of such warfare, hence Carthage's compliance, even if the citizens wept. Rome was less pernikerty, and did not mind indulging in further wars with already humiliated enemies, but only if they were either near or especially rich. The wars fought after 188, therefore, tended to be foisted on the victims by Roman paranoia or greed. And the Roman army was usually in good enough fettle, what with its continuous warfare on the Italian and Spanish frontiers, that there was little difficulty in winning, other than those problems resulting from the incompetence of the commanders.

At the end of the war with Antiochos it is explicitly stated that fifty Roman ships were laid up.[5] This was the Roman fleet which was kept for emergencies, but it was not employed again until the war against Perseus of Macedon which began in 171. Squadrons were activated for lesser conflicts – twenty ships were used against pirates in Liguria and Illyria in 181–176,[6] and ten against Sardinia in 177.[7] But in 171 only thirty-eight of the fifty at Rome were seaworthy, and they had to be supplemented by twelve from Sicily – which implies that there was a regular naval command in Sicily as well.

These were conflicts with single enemies, or at least in a single part of the world. Between 149 and 146, however, Rome was involved in wars in both Greece and Africa. Until then Rome had felt the need to mobilize only a relatively small fleet: fifty ships and a number of naval allies for the war against Perseus,[8] and smaller squadrons elsewhere. In 149–146, when wars in Africa and Greece had to be fought simultaneously, fifty quinqueremes and a hundred *hemioliai* were sent against Carthage, but only transports were needed to take the army to Greece and Macedon,[9] for neither of the enemies there, the Akhaian League and Macedon, had any naval power. A decade later, when war came in Asia over the future of the Attalid kingdom, the sea power of the Attalid kingdom was not apparently available to those who disputed the Roman takeover, and it appears that local allied forces in the Aegean area sufficed. The pretender Aristonikos failed to capture Elaia, the Attalid naval base, and those ships he did gather were soon defeated by the city fleet of Ephesos.[10]

In between these wars most of Rome's ships were laid up, so that they were available for use if required. In 181 and 177, ships were needed against Ligurian and Illyrian pirates. Especially in 181 this provoked the Senate into more extensive measures than usual. Twenty ships were to be used, and since there were simultaneous piracy threats in two areas, Liguria and Illyria, the old office of *duumvir navales* was revived. Between them the two men were to supervise the whole coast from the borders with Massalia to Barium in Apulia, the dividing line being Cape Minerva, which is probably the Sorrento Peninsula.[11] The man given the northern assignment, C. Matienus, captured thirty-two Ligurian ships when the Apuani were compelled to surrender.[12] The other commander, C. Lucretius Gallus, appears not to have had to do anything, but he was elected praetor in 171 and given command of the fleet sent against Perseus of Macedon,[13] though the naval war was actually against King Genthius of Illyria in the main – no doubt Lucretius' experience in those seas was a factor in his election.

More duumvirs were elected in 178, and both went to the Adriatic. Matienus' work had largely subdued the Ligurian piracy in 181, and now it was the turn of the Illyrians to be suppressed. The two men divided the Adriatic coast at Ancona, one, C. Furius, having responsibility as far north as Aquileia, the other, L. Cornelius, as far south as Tarentum. Each man commanded ten ships, and those in the north were used to escort transports for a war against the Histrians.[14]

The only permanently embodied squadron seems to have been one in Sicilian waters, no doubt based at Lilybaion or Syracuse, which was called on in 171 when the main fleet was found to need replenishment.[15] The fifty ships launched against Perseus in 171 had been kept available for seventeen years. In fact it turned out that only thirty-eight of those at Rome were found to be seaworthy; the total was made up with twelve from Sicily. This is the evidence for a squadron of ships having been maintained in times of peace in the island, though it is possible that the ships were laid up there as well. The Senate had authorized the building of an extra thirty quinqueremes during the war with Antiochos, in 191, which should still have been usable. This would seem to be the total of Rome's navy by 171.

The fifty ships launched to deal with Carthage in 149 are not specified as either old or new, but one would expect at least some of them to be new, since those of 171, already twenty years old by then, if not more, are unlikely still to have been available. *Hemioliai*, a smaller galley with one and a half banks of oars, appear here in the Roman fleet for the first time,[16] so these were probably newly built, unless they had been acquired from the naval allies. Their use was a delayed reaction by the greatest naval power to the demonstration of the effectiveness of smaller warships such as the *lembos* by Philip and Perseus.

In the Third Romano–Macedonian War (171–167), much of Rome's naval force was contributed by the allies; their ships consisted of smaller vessels rather than the quinqueremes the Senate favoured. Before the army and the fleet set out from Italy in 171, Rhodes had been contacted and had offered to assist with forty ships, though most of this offer was not in the event taken up;[17] Rome did not wish to be too much obligated to any ally. In Italy, Rhegion, Lokroi and 'Uria' (probably a mistake for Thurioi) contributed seven ships between them, the island-city of Issa twelve ships, and in Illyria the city of Dyrrhachion produced ten *lemboi*.[18] King Genthius of Illyria, who was already regarded with some suspicion at Rome, proved to have no less than fifty-four *lemboi* ready for use when he was visited by C. Lucretius Gallus; the size of this force only confirmed Roman suspicions of the king; nevertheless these ships were all blandly accepted for the Roman service; if they were being used by Rome, Genthius could not use them against Rome.[19] Reinforcements were sent to Issa, where C. Furius, formerly a *duumvir* in the Adriatic, was now legate on the island in command of two ships of that city; 2,000 men were sent in eight ships (or eighteen, which is more likely to have been able to carry so many). The eight ships were from Brundisium, and were clearly sent to deter Illyrian raids.[20]

Most of the Roman fleet was taken round to the Aegean, but there was little in the way of naval warfare for it to do. The first praetor in charge instead got himself involved in the land campaign, presumably so as to be able to boast of a visible achievement, and acquire loot. Later a number of local allies around the Aegean provided more ships: Herakleia Pontike contributed two, Chalkedon four, and Samos four; two quinqueremes came from Carthage, and six of the offered Rhodian ships were taken up, but these were at once sent back since warfare at sea did not happen.[21] The naval praetor had taken forty Roman ships to the east in 171 (not the whole fifty which had been launched). At one point in the war King Eumenes came across the Aegean with twenty ships, and king Prusias of Bithynia sent five.[22]

These contributions from eastern allies were hardly needed. Indeed Eumenes' ships operated independently of the Roman fleet and were on an expedition of Eumenes' own. The purpose of their being utilized is not stated, though some Rhodian and Attalid ships are mentioned as blockading a Macedonian supply convoy at Tenedos.[23] Since the ships were not really needed, it follows that these contributions had a different purpose, which was essentially political. Each small contribution was a sign of the home city's subordination to Rome. The Carthaginians had provided ships both to the war with Antiochos and that with Perseus, as did Rhodes. By offering to help before a direct request came in they might hope to score points at Rome; other cities no doubt took note. Collectively these contingents were a display of the extent of Rome's power, a demonstration to Perseus above all of his isolation. This did not stop the Roman attitude to these allies souring throughout the war.

The allies' attitude to Rome also soured, and one of the reasons was the behaviour of successive naval praetors. Lucretius was accused of theft and oppression at Chalkis, his main naval headquarters in the Aegean during his time in command. His successor, L. Hortensius, went one stage further and seized control of the city of Abdera on the north coast of the Aegean, a free ally of Rome. He plundered the city, executed the leading men, and sold others as slaves. As the news of this spread, other places which were approached by the fleet under his command simply shut their gates and refused him entry. Emathia, Amphipolis, Ainos and Maroneia, all neighbours of Abdera, are mentioned as doing so.[24]

The Roman conduct of the war was sloppy and inefficient, as well as being at times more dangerous to Rome's allies than to her enemies. The commanders were careless and uniformly unsuccessful until L. Aemilius Paullus was appointed in 168. An official investigation instigated by him reported that the fleet was undermanned, partly because many men had died of disease, partly because others, particularly men contributed by the Sicilian naval allies, had gone home; none of the men still on duty had been paid, and their clothing was in rags. No less than 5,000 extra men were levied from the naval allies to make up the deficiency; even more reinforcements went to the army.[25] This display of senatorial favouritism, greed, cupidity and incompetence annoyed the Greek allies, whose complaints in turn annoyed the Romans. The result was that the local allies in the Aegean area were now being regarded at Rome with the same suspicion as Genthius; their lack of enthusiasm – Eumenes had gone home with his ships – was a reflection of the bad, corrupt, and stupid Roman military performance. The allies could not be expected to assist with any real enthusiasm when Rome was operating so badly.

The enemy, King Perseus of Macedon, under pressure in the land war, even though he was able to prevent an invasion of its home territory, was unable to muster much support. The only open assistance came from Genthius, who was able to send eighty *lemboi* on a raid against the cities of the Illyrian province. He was, however, rapidly suppressed and his ships defeated.[26] (Ten Roman ships had been left in Italy somewhere, probably Brundisium; in view of Roman suspicions about Genthius from the start, they were presumably available to be used against him, just in case he chose to fight.)

Perseus, no doubt aware of the failure of the Roman fleet to be effective, launched his own squadron of *lemboi*. Under the command of Antenor, forty of these ships went on a disruptive Aegean cruise. The blockade of Tenedos by Rhodian and Attalid warships was broken, and a convoy of grain ships which had been held in harbour there was sent on to Macedon; the Hellespont was threatened; a convoy of reinforcements for the Attalid troops in Greece was intercepted and destroyed. In all these cases surprise was total. There had clearly been no Macedonian activity at sea until this raid. Antenor made the holy island of Delos his base for a series of raids by small groups of ships throughout the 'Rhodian' Kyklades, even though Delos was nominally neutral, because of its sanctity. These raids were ineffectively countered by Roman ships, which, as usual in such warfare, always arrived too late. This was disruptive, but it was not something which would seriously disturb the Roman war effort in the long term; it was in fact largely directed at the allies, not the Romans, no doubt deliberately.[27]

In 168 two efficient and tried commanders came to take over the Roman forces: L. Aemilius Paullus, in his second consulship, on land, and Cn. Octavius for the fleet. Octavius moved the headquarters of the fleet from Chalkis to Oreus, a day's sail closer to the enemy coast,[28] and used it, now fully manned, to menace the royal dockyards in Kassandreia and Thessalonica and Dion, and to threaten landings along the Macedonian coast,[29] but not, presumably, at the free Roman allies. This forced Perseus to disperse part of his army to provide coastal protection, and still more of his forces were with the raiding *lemboi*. This was a significant assistance for Paullus' land campaign, and it was also a sensible use of the fleet, which was effectively unopposed on the water. Paullus' land campaign culminated in a great defeat for Perseus at Pydna in 167. The fleet was offshore at the time of the battle, and did its part in cutting down many of the fugitives.[30] Octavius' work was so effective he received a naval triumph; he was consul only two years later.

But it was Paullus who took back to Rome the old sixteen of King Philip and his predecessors, having himself rowed majestically up the Tiber amid the cheers of the people;[31] the ship was housed in a special shed, probably until it fell apart – the shed seems to have been available to house Carthaginian hostages less than twenty years later.[32] In a way this might be seen as a symbol of the transfer of admiralty from Greeks to Romans, though the actual change had happened fifty years earlier.

It is worth noting that the scattered notices of allied contributions to the Roman wars show that, besides the fleets of the Great Powers – or the single

Great Power now – there was also a fair number of warships in the main port cities around the Mediterranean. Only in the notices of these wars are they mentioned. Warships are noted at Carthage, Naples, Rhegion, Lokroi, probably Thurioi, Issa, Dyrrhachion, Athens, Ephesos, Samos, Chalkedon, and Herakleia Pontike. And that is not counting the substantial fleets of Rhodes and the Attalid kingdom, nor the fleets put to use by Kings Genthius and Perseus, and by the Ligurians. Massalia had a local fleet as well. There were undoubtedly others (some will be noted later), for it would seem that virtually every port city had a small number of such ships, and it was the crews of these places which were called up by Rome to serve in time to war. Rome might insist on enemy ships being destroyed or burnt or surrendered as part of a peace agreement, but later enforcement was clearly lax, and there were always plenty of other ships in other cities. Rome's naval supremacy stands out in the wars, but in the intervals of peace it was local power which counted.

While Rome fought in Africa and Greece, another war had broken out in the eastern Mediterranean between the Seleukid and Ptolemaic kingdoms. The Sixth Syrian War had its origins partly in the Ptolemaic wish to recover the lost Syrian territories, and partly in the instability of the Ptolemaic government, which was under the control of a pair of regents for child monarchs whose position was under threat. It was mainly a land war, but Antiochos IV, the Seleukid king, was able to capture Cyprus late in the war, which presupposes a fleet of sorts,[33] though he was compelled by Roman intervention to return the island. By that time the Battle of Pydna had been fought and won; the Roman fleet (and army) was free to sail east if necessary, though there is no suggestion that this was in any way part of the Roman plan. Antiochos gave in to the Roman demand that he make peace with Ptolemy, at once. His object in continuing the war had been to safeguard his possession of Koile Syria and to weaken Egypt, and these had both been achieved. Roman intervention in effect had got him out of a war that he knew he could not win in any other way. He also knew that Rome was not interested in any further involvement in the east.

The corrupt and incompetent performance of the Roman commanders in the first three years of the Macedonian war had led Rhodes and Eumenes to consider changing sides, or at least to seem to be considering it, or to becoming neutral. Rome reacted by developing suspicions of all the eastern states and rulers, suspicions based essentially on contempt for their weakness. All of them, from Carthage to Syria, were systematically further undermined and weakened. Eumenes was insulted by a hastily devised measure at Rome stating that kings could not visit Italy, produced just when he was about to visit, and by an intervention which deprived him of the fruits of victory over the Galatians in central Asia Minor. Rhodes was compelled to give up its possessions in Karia and Lykia, and was weakened commercially when Delos was given back to Athens and designated as a free port. The Macedonians under Antenor had used the holy, neutral island as a base for raiding in the region during the war; the Roman measures completed its desecration by making it a vibrant commercial entrepot. The fate of the Island League is not known, nor whether Rhodes continued to be its patron, but when Rome detached Delos from the league it would obviously reduce Rhodian influence and authority among the rest of the islands. Macedon had been broken up into four weak republics, none of which had any sea capability. Akhaia was virtually harmless in the face of any Roman power, but it was still punished by having 1,000 of its leading citizens removed to Italy as 'hostages'. The league had never had more than a few ships.

In terms of sea power, this period of warfare eliminated any possible challengers to Rome even among the smaller powers. With Rhodes and Eumenes deliberately weakened, no Greek state could launch more than a few triremes. Neither the Seleukids nor the Ptolemies had the ships capable of challenging Rome, though they both had small fleets, and Ptolemy still held Cyprus, Itanos, Thera, and Methana. Given Rome's suspicions, this implies there was now a complete lack of any real Ptolemaic sea power. The mutual antipathy of Seleukids and Ptolemies neutralized them so far as active enmity towards Rome was concerned. And when both kingdoms fell into civil warfare in the late 160s, the Roman Senate was more than pleased. Cn. Octavius, the former fleet commander in the Aegean, now of consular rank, went to Antioch as legate during the minority of Antiochos V. The regent Lysias was quite unable to resist him when he resurrected the terms of the Treaty of Apamea, by then a quarter of a century old and never enforced until this time, and insisted on the destruction of ships and elephants.[34] In other words, the Seleukid kings had either kept the ships which should have been destroyed, or had built others (just as Antiochos III had been compelled to give up his elephants, but his successors had acquired more). Octavius was assassinated by an aggrieved citizen in revenge for his actions.[35] This was widely approved by many other Seleukid subjects, who were incensed by the sufferings of the hamstrung elephants as they died. The Senate did nothing, partly because of the sheer distance of the Seleukid kingdom from Roman territory, and partly perhaps because Octavius had been in the wrong – though being in the wrong has never stopped any Great Power from exerting itself when it felt it necessary.

Rome was therefore attempting to control the Mediterranean by influence, threat, and opportunistic destruction, all conducted at long range. But it was doing so without the clear exercise of power, which in the Mediterranean needed warships. After each war most, if not all, of the Roman warships were laid up. That is, Rome was relying on the memory of its earlier exercises of power and brutality to maintain its political influence. Yet the longer this continued, the fainter became the memories of Roman military and naval success, and the likelier it was that other states would defy the Great Power.

By 150 this approach was failing; in the next five years there was defiance from Carthage, though the city did everything possible to avoid being attacked by Rome, from Macedon, which welcomed a pretender to the vacant throne in order for the kingdom to be reconstituted, and even from the Akhaian League, which was effectively unarmed. While Rome dealt with these crises, by invasion and frightfulness, the Ptolemaic king Ptolemy VI reconquered the lost province of Koile Syria, only to be killed in the final battle. One result of the overall crisis was that his successor, Ptolemy VIII, evacuated the three Ptolemaic posts in the Aegean. Rome now finally accepted the logic of empire and took direct control of Carthage, Macedon, and Greece. For Ptolemy it was clearly best to keep a clear physical distance between Rome and Ptolemaic territory. Ptolemy VI had in fact used the Ptolemaic fleet in his Syrian campaigns, but it was of no real size, and certainly not a threat to Roman hegemony, but with Rome in a mood of anger it was best to keep one's distance.

Rome had mobilized the fleet to attack Carthage, but had not apparently needed to send ships to Macedonia to assist in the suppression of Andriskos. In the war in Greece, some communities were said to be unwilling to join the general Akhaian muster for fear of the Roman fleet,[36] though that is not proof of the fleet's proximity. Few ships were needed in the Greek wars.

Carthage, on the other hand, possessed some ships of war when the crisis came to a head, which the Romans required to be destroyed as one of the conditions they imposed for not attacking the city;[37] whether the city complied is not known. The Carthaginian army was also to be disbanded as part of the conditions, and that did not happen, so perhaps the ships were also kept. When the Roman expeditionary force sailed, the Carthaginians did not attempt to intercept it, but that is not an indication of the absence of a Carthaginian fleet, since the Roman fleet was clearly far too strong. Driven to war to try to preserve their city, the Carthaginians stood a siege, displaying the same sort of persistence, courage, and ingenuity as their Tyrian relatives had against Alexander. The Roman fleet was threatened at one point with destruction by a fire-ship attack,[38] forcing the consul to fortify his camp, and then to move it; the original palisade had to be replaced with a wall.

The allies were requested to contribute to the enterprise, though only one is named. Side, a city in Pamphylia, sent at least five ships, which distinguished themselves in the fighting.[39] Others seem to have contributed as well, but with wars in Greece and Syria at the same time, there was not much available.

The Roman fleet was based at the city of Utica, which had been an ally of Carthage rather than a subject. A fairly loose blockade of the city of Carthage was imposed, but supplies could get in when the wind was favourable. When he took command, in 147, after two years of indecisive fighting, the consul Scipio Aemilianus first cut the city off by land by digging a ditch across the isthmus, then began an embankment designed to block access to the harbour. The

Carthaginians replied by digging a new harbour entrance out of which came a fleet of ships, said to be fifty triremes, which they had built using timbers already available in the city, and many smaller ships of various types came out at the same time.[40] It seems unlikely that a total of fifty ships could have been built quite so quickly, and another comment by Appian, our source for these events, mentions that quinqueremes had been built as well. Some of these ships were probably those which the Romans had demanded to be destroyed earlier. (On the other hand large fleets could be conjured up quickly when the materials, the finance, and the will and skills were present – note Antigonos in Phoenicia in 315/314, or, indeed, Rome several times in the first war with Carthage.)

When the ships came out they provided a salutary shock to the Roman fleet, but they merely paraded about rather than attacking the Roman ships, so providing the Romans with a warning of what was to come, and when they came out again, therefore, the Romans were prepared. The fight was about equal until the Carthaginians retired, when the harbour entrance became jammed with ships, and the larger vessels had to take station along an adjacent quay. They successfully defended themselves until the ships from Side, fighting as allies of the Romans, worked out a way to attack them and then get away without damage, by laying out ropes with which they could pull themselves away without becoming vulnerable. When the rest of the Roman fleet followed the example of the Sidetans the battle finally went the Roman way.[41]

This was the only sea fighting involved in this war, though this was in part because, now warned, Scipio made the quay where the fighting had been so intense his next target. It took some severe fighting to secure control of it, but when he did, the exit from the harbour was blocked. The city was finally cut off, and its supplies were already short. But it was by assault rather than starvation that it was finally taken, and ironically, it was through the naval harbour that the Romans first broke in.

The city burned during the assault, which accomplished the Roman purpose in destroying it. To some extent the resistance the Carthaginians put up was a testimony to the insistence by the Senate that it be destroyed, and the ability of the Carthaginians to conjure up a fleet, even one which was beaten in the first battle, gives some idea of the basis of the Romans' fears. Nevertheless Carthage had not been a serious threat to Rome; its destruction, like that of Corinth which happened about the same time, was undoubtedly a crime – but winners are not, of course, called to account, except by historians.

The annexation of the conquered lands further removed any threat to Italy by sea. The Roman fleet which had been used at Carthage was laid up once more, and no Roman ships appear again for some decades. Even in the war for Asia in 133–129 no fleet is noted, though admittedly there are few and poor sources for this war. This was all perfectly reasonable, for there was no navy capable of mounting an attack, just as no state was strong enough to do so either.

Chapter 13

Pirates

Piracy was endemic in the Mediterranean from the time ships sailed on its waters. It remained endemic until the nineteenth century AD. It was something that merchants and sailors automatically took into their calculations in planning voyages, buying insurance, manning their ships, and deciding when and where to sail. It was a practice no power ever succeeded in extirpating until the fast steam-driven vessels of the Royal Navy came to dominate the sea. Other navies at times restricted the pirates' activities, though usually only in parts of the sea. The Hellenistic world put up with it, as did all other periods.

For about half a century during the late Hellenistic period, however, piracy grew to be such a plague that it called forth, eventually, a major effort of suppression. That it could be largely suppressed by that major effort, and relatively easily, is a sign that it had been almost deliberately allowed to grow. For piracy existed and flourished because others, besides the pirates, profited from it. It was, that is, a combination of the normal crimes of theft and violence which just happened to need an unusually large capital investment.

The term 'piracy' is rarely applied by pirates to themselves and their activities. It is a description of the activities of one's enemies. It is therefore not necessarily an accurate characterization of that activity. The Phoenicians were regularly described as pirates by the Greeks, but most of what we know of them suggests trade with or without violence. In the Hellenistic period the Aitolians are regularly described as indulging in piracy as a sort of state policy, but all such descriptions come from their enemies and are not to be trusted.[1] An investigation shows that there were few or no cases which can be really described as piracy – though there were individual Aitolians who were pirates, such as Dikaiarchos, one of Philip V's employees, which is not the same thing. Similarly, some of the men who fought at sea for Antiochos III were called pirates by their enemies, and it is their characterization which has entered the sources, yet they seem in fact to be just sailors. It is thus necessary to be wary.

At various times, states with considerable sea power had interested themselves in controlling the problem. Athens did so in the fifth and fourth centuries BC, at least in the Aegean Sea and the waters around Greece. This, of course, was largely a matter of self-interest: Athens depended on the import of grain from the Black Sea, and it had to justify its taxation of its imperial subjects; piracy, if it became too obvious and too disruptive, would interfere with both of these. Quite probably Carthage in western waters was similarly active in its suppression, though little is known of Carthaginian maritime policy.

The islands and intricate coastline of the Aegean Sea were the most prolific breeding ground for piracy, hence Athens' concerns. Just as effective in producing pirates was the much fragmented political state of the area. This permitted pirates to sell their loot and captives in places from which neither they nor their victims came, and the existence of such markets was one of the elements which fuelled the practice.

In the early Hellenistic period one of the main sources of pirates seems to have been Italy. There are several reports of 'Tyrrhenian' pirates, that is, pirate ships sailing out of ports on the Tyrrhenian coast of Italy, active as far east as the Aegean and, no doubt, elsewhere. Antium, for example, was a notorious pirate centre until captured by the Romans, and even after that its men were still liable to become pirates.[2] Athens was busy founding a colony on the Adriatic coast explicitly to combat the danger from such pirates when the news of Alexander's death arrived;[3] it may have failed when Athens succumbed to Antipater and its navy faded away – its location is not known. Agathokles in Syracuse hired a squadron of Etruscan pirates, who were of particular use in the decisive fight for the city.[4] Demetrios Poliorketes is said to have sent back some 'Tyrrhenian' pirates to Rome, with a polite complaint.[5] Aratos of Sikyon, trying to evade Macedonian interception on his way to appeal for help from Egypt, was almost captured in the Adriatic by a 'Roman' pirate.[6] After Rome gained a more certain grip on the whole peninsula, however, references to 'Tyrrhenian' pirates tended to die out.

Several regions of Greece were equally notorious as havens and sources of piracy, and Crete was always regarded as a major source. When a major sea power existed, such activities were kept within bounds, as with the Athenian empire, but large scale warfare at sea tended to encourage private maritime enterprise, simply because the warship fleets were preoccupied with each other. At the great siege of Rhodes in 305, Demetrios was supported by a thousand 'pirate' ships, waiting the opportunity to loot the island.[7] Whether these really were pirates is an open question. The source for the description is a Rhodian one, and one does not expect the victim to be polite about the enemy. Some of those thousand ships will have been merchants bringing supplies to Demetrios' forces, while others were waiting the chance to buy the loot he had acquired. This is the same problem which exists with references to Aitolian piracy, for which the source is Polybios, who had a political motive (he was from Akhaia, which had suffered from Aitolian enmity) in painting his country's enemy in the darkest colours. When provided with such a source, it is not difficult to interpret other sources in its light. Ignoring it, they become less like criminal activities and more like normal warlike relationships.

So it is necessary to keep a sense of proportion. 'Pirate' was an easy insult to brandish at an enemy if the sea was involved; it is best to make sure that no alternative motive is involved before accepting the insult as an accurate characterization. Not all that is labelled 'piracy' was actually that; some of it, perhaps a lot of it, was the result of enmity for other reasons. And it is necessary to recall that pirates who steal have to dispose of their loot. The obvious places to do so are markets. So one person's pirate is another person's merchant, and what are stolen goods to the victim is a good bargain to the purchaser. In a way, of course, piracy is a form of trade.

Yet it is also a restraint of trade, and since no one likes to be robbed, anti-pirate precautions will put up the price of goods. The precautions might include buying insurance, or buying weapons, hiring extra crewmen so as to be able to resist attack, using a larger ship, or putting into port every night and so travelling more slowly and having to pay harbouring fees. All of these measures increased merchants' costs, either directly or by slowing down the movement of goods. Piracy was, therefore, necessarily a criminal activity, one on which most governments frowned, and one which, if they could, they suppressed.

But piracy was also a profession whose practitioners were likely to be hired by any government requiring military or naval manpower urgently, such as in a war. Antigonos Gonatas hired an Aitolian pirate called Ameinias during his wars to acquire Macedon;[8] Antiochos III hired men whom his enemies called pirates – 'Nikadoros the *archipiratus*' – during his war with Rome.[9] These men were usually paid off when the immediate task had been accomplished; that is, during the war they were actually mercenaries, and perhaps they became pirates at other times. Crete, of course, was a major source of both pirates and mercenaries, who were probably the same men.

The practice of piracy, therefore, clearly shaded off in various directions into other professions, and those other professions may well have been quite respectable, or at least on the acceptable side of the divide between criminality and non-criminality. On one side they could be seen as merchants, or as mercenaries, or no doubt as peaceful sailors. Opportunity was all: if it was thought more profitable to steal, they became pirates. On the other hand, if there was a well-policed sea, they were more likely to be content with the profits of trade or mercenary pay.

The crucial element therefore was the vigilance of governments. This was applied in two ways. The obvious one was that the state from which the pirates operated should control them. This was the approach the Roman Senate applied for a long time, and seems to have been successful in suppressing piracy among its own subjects. The other approach was to suppress piratical activity at sea by dominating the sea with a fast-reacting fleet of ships which patrolled incessantly, which convoyed merchant ships, and kept a close eye on the places from which the pirates came. Athens had controlled piracy in the Aegean in this way in the fifth and fourth centuries, and the Ptolemaic navy's presence in the Aegean in the third century had a similar effect, for a time. Also in the third century, Rhodes gained a reputation for hostility to piracy which lasted into the second century.

These two were the approaches which worked, but they had to operate in tandem. The vigilance of states by controlling their subjects was the most effective way. Rome, for example, sent expeditions against the Illyrians in 229 and 219, after which a large part of Illyria came under Roman control, and much of the rest of the area was within a kingdom whose kings suppressed the piracy which many of their subjects wished to practice. This worked under Kings Skerdilaidas and Pleuratus, but when Genthius took power with a different policy, piracy began again, or rather what its victims and enemies called piracy; since it was the official policy of King Genthius it was in fact warfare. Rome was able, by its proximity, and by Genthius' political miscalculation during the war between Rome and Perseus of Macedon, to re-establish some control.

The other method, suppressing pirates by sea patrols, was a task which required constant naval activity, a constant naval presence where it was liable to break out, and that was expensive. Suppressing piracy was therefore something only a prosperous naval power, whose merchants were making money and who were willing to pay taxes, could undertake. It also took a considerable and persistent determination to undertake the task. In the second half of the second century BC no power would or could do this.

Rome laid up its fleet after every war, but quinqueremes were too heavy and ponderous to be really useful in the face of the nimble craft used by pirates – in the Adriatic these were, of course, *lemboi*. This is probably the reason individual cities tended to maintain a few triremes, as a deterrent and perhaps as a means of escorting their own ships. Rome had, of course, a perfectly serviceable institution, flexible and adaptable, in the *duumvir navales*. These could be deployed when needed, as they were originally along the Tyrrhenian coast, then from Massalia to Barium, then in the Adriatic. The career of C. Furius, *duumvir* in 178 and legate in Issa in the war against Genthius, illustrates the possibilities. But this magistracy was discarded, so it seems, after about 170, leaving the policing of the seas to local initiative.

The goods which pirates sought were, above all, people. By capturing people by violence the victims became slaves and could be sold in a slave market. These 'goods' would be acquired by capturing ships at sea, or, more likely, by landing at an unguarded coast or an island and kidnapping anyone who could be found. They were then carried off to another place to be sold. This was a favourite scenario for Hellenistic novelists, the victim usually being a beautiful maiden.

Despite their anti-pirate reputation, the Rhodians ran an important slave market on their island, and many of those sold in it were probably victims of pirate kidnappings.[10] When Rome made Delos into a free port after 167, that holy island became a major slave market as well, causing much scandalized comment; but it still continued.[11] This was not, of course, the intention, but trade of any sort will flow to the least cumbersome market.

That is to say, the suppressors of piracy were also the pirates' customers, and it is clear from some inscriptions that Rhodes was inimical only to piracy which damaged its own trade; pirates who damaged Rhodes' own enemies were not a problem.[12] About the only restraint was that victims could not be sold in the slave market of their own city, but there were plenty of other available places. Ransom was an alternative, if it could be arranged, and this was probably the preferred alternative by all sides.

The emergence of Delos as a major slave market was partly the result of its free port status, but also that it had been given to Athens by Rome, for this put the island under Rome's distant protection. This was quite reasonable, since it was Rome and Italy which were the main destinations for the slaves who passed through its market. The transfer of much of the wealth of the eastern Mediterranean to Rome as a result of the Roman victories was the fuel for this. It was also the Roman victories over the other sea powers, and the concomitant destruction, or at best restriction in size, of rival navies, which opened up the seas to the pirates. So in the second half of the second century there was less policing of the seas, except on a very local scale, and at the same time there was a hugely increased market for their goods.

The crises of the 140s began the process. Rome had already punished Rhodes and the Attalid kingdom after the Macedonian wars for their display of a certain independence of attitude, and they necessarily now had to be careful about exercising any power at all. In the 140s a complex civil war racked the Seleukid kingdom, and reduced its authority over its outlying ports: Palestine was in rebellion, for example – Joppa became a pirate port – and Iran was a battleground between the Seleukids and the Parthians. Wars always fuelled the slave trade because armies sold captives, while the state governments, which might control and limit the exportation of their people, had collapsed, so permitting piratical activity and failing to protect their territories.

The particular peripheral territory which became the most notable base for piracy was Kilikia. This was a province in two, or perhaps three, parts. Smooth Kilikia was to the east, separated from Syria by the Amanus mountain range, rich and fertile and well-watered, a land of cities which was generally under constant Seleukid control. Inland were the Taurus Mountains, forming Kilikia's northern boundary, never under anyone's control. These hills were the homes of people who could turn raiders if the lowlanders' vigilance relaxed. The area became the home of at least two Seleukid pretenders in the 150s, who conducted raids into the lowlands.

The western part was Kilikia Tracheia, Rough Kilikia, a land of hills, valleys, steep slopes, and narrow tracks, a land of villages, small towns, and many ports, which had been rarely under any government's control. Like the Taurus Mountains it had never really been part of any of the Hellenistic kingdoms, who had restricted themselves to possessing the coastal cities, though powerful neighbouring states – the Seleukids, the Ptolemies, and before them the Akhaimenid empire – could suppress the most egregious manifestations of piracy. The easiest way to travel was by sea; the coasts were the homes of plenty of skilled sailors. Antiochos III gained control of the coast in his campaign of 197, but lost it again at the Treaty of Apamea in 188. He kept Smooth Kilikia, but Kilikia Tracheia ceased to be part of any state, and the Seleukids were forbidden to send warships beyond the western boundary of Smooth Kilikia.

The turmoil in Syria in the 140s spilled over into Kilikia Tracheia, particularly when one of the pretenders to the kingship, Diodotos of Kasiana, who took the throne name Tryphon, hired local sailors as privateers.[13] Tryphon failed in his bid for the kingship, and Syria calmed down for a time, but it seems that the Kilikians had acquired a taste for piracy. Little is known of events in the Kilikian area for a time after Tryphon (who died in Syria in 138), but mutual enmity between the several rulers in Syria, Cyprus, and Egypt, and their failure to maintain their navies, permitted the pirates to grow in activity. Strabo the geographer named Korakesion as their 'headquarters', though what this means for a set of people unlikely ever to cooperate with each other for long is uncertain.[14] Side, a little further west, is also stated to be an important slave market used by the pirates, as are other ports in Pamphylia, but the main market was at Delos, capable of processing 10,000 slaves a day.[15]

Rome's main involvement in all this was as the main market for the slaves. But it was also involved, less directly, as the one power in the Mediterranean which had the political and naval strength to tackle piracy. No other state had the ships, and the treatment of Rhodes and Carthage and the eastern kingdoms by Rome showed that no other state could be permitted to develop a navy large enough to deal with the problem – for if it did, it would have a navy large enough to threaten, or at least cause trouble for, Rome. But Rome itself was unwilling to intervene, and blamed the local powers for not controlling their territories properly. This may have been the conclusion of Scipio Aemilianus, who made an official tour of the east in the late 140s,[16] but it is more likely to be a later conclusion, based on general ignorance rather than on a close inspection.

Rome's reaction to the growing pirate problem may have in part been one which betrayed indifference as to the condition of the eastern Mediterranean, and a certain impatience with those whom it hoped would tackle the problem themselves, but it was not the result of a failure to know what to do. In the second century at least three pirate sources had been identified and dealt with by the Roman government efficiently and, so far as can be seen, permanently. All three, Ligurians, Illyrians, and Balearic Islanders, were tackled by a combination of naval patrols and a land campaign. For, like Kilikia, all three areas of piracy had been lands which did not have a central government to control the population. Further, during the period when Kilikian piracy was becoming a growing problem, the Balearic Islands were actually brought under control by a Roman mixture of ruthlessness and efficiency.

That it was not merely an eastern Mediterranean phenomenon is indicated by the need felt in the late 120s to conquer the Balearics, in part because they were sources of pirates.[17] This was also a period when southern Gaul was organized into the province of Gallia Narbonensis, which safeguarded the land route to Spain; controlling the Balearics did the same for the sea route along the southern Gallic coast. The impulse for this can perhaps be put down to the need to protect the increasing amount of seaborne traffic moving between Italy and Spain. Complaints by Massalia, whose trade had perhaps suffered most, may also have been involved. Massalia had also complained in the past about pirates operating out of Liguria, geographically a very similar land to Kilikia Tracheia – mountains coming down to the sea, with consequently many small harbours – which had already been addressed in the 170s. The Ligurian piracy problem had partly been the result of the Roman military campaigns of conquest in northern Italy. For a time it was left to Massalia to deal with the problem, but by 181 Massalia could no longer cope. For the next five years, as noted in the previous chapter, *duumvir navales* were appointed. This was a clear case, for those Romans capable of understanding, where the Roman policy of suppressing local navies simply left the work to be done by the Romans themselves – something they were generally reluctant to do.

So when the Balearics proved to be a pirate base, it was simple enough for Rome to deal with it. The methods were known. The islands were conquered with some thoroughness by Q. Caecilius Metellus, consul for 123, and a colony of Roman citizens was planted there. Piracy is not mentioned again in reference to the Balearics, though this is not to say it had wholly disappeared, only that it was no longer unpleasant enough to trigger a Roman response.

The Romans therefore fully understood that piracy at sea could not be suppressed by action at sea alone. The only secure way to proceed was to gain control of the pirates' land bases. This presumably lay behind the regular blaming of eastern governments for letting pirates operate. But many of the pirates' bases were actually outside the territories of any of the major governments, and a numerous enough pirate community could dominate a small port-city without much difficulty. By blaming governments the Romans were in fact pointing to the correct way to deal with them, by exerting control. On the other hand, the Senate was all too ready to deprive any state of its conquests if appealed to by the victims – Rhodes was the prime example – so one of the main incentives for taking military action had been removed.

Complaints to Rome increased, therefore, until the Senate finally took some action, though it was basically an adaptation of the idea of local responsibility. M. Antonius, praetor in 102, was sent to take command against pirates in southern Asia Minor, using ships and men gathered from anyone who had them or was concerned.[18] A fleet of some sort was, so it seems from an undated inscription, brought across the Corinthian Isthmus by Antonius' propraetor, A. Hirrus, who then spent some time at Athens waiting for improved weather.[19] It may be assumed that the ships he brought were from Italy, but how many there were, and what type, is not known. Rome itself had no navy by this time, but the Italian and Illyrian naval allies could have produced some ships. The Byzantines certainly contributed ships,[20] so did the Rhodians; possibly Athens added some, maybe Tenos, and some of the cities of Lykia. The sources for this expedition are scattered and discontinuous, so it is likely that other cities contributed ships as well. Nor is it clear what Antonius actually did. Livy says he pursued the pirates into Kilikia, and the Isthmus inscription says he was headed for Side. This would be a reasonable base for a campaign into Kilikia Tracheia, and since he had a fleet it is probable that the campaign was along the coast, dealing with each of the towns and cities as he reached them, using such soldiers as were in the ships and those he could recruit locally. He was awarded a triumph on his return to Rome, so his forces clearly killed a reasonable number of people. But piracy was not seriously reduced.

It was, in fact, in the generation following Antonius' expedition that piracy grew to be a major menace to all the societies around the Mediterranean, as the pirate numbers grew, and Rome collapsed into civil warfare. The Senate produced a new law in 100 BC designating the pirates as enemies and instituting a new province of Kilikia, but neither had any effect on the pirates.[21] They were enemies anyway, and the new province was mainly an area of command of an uncertain extent, not a territory which Rome had taken into its empire.

So the pirates continued and expanded their activities. It is in this period that Korakesion was a pirate stronghold, and the market at Side became an important pirate outlet. The Roman measures against the pirates in the east had in fact been wholly unavailing, as had those taken locally. Once again the local rulers, who had other priorities, were exhorted to see that their territories were not used as pirate bases; and once again the Senate, like those it exhorted, also had other priorities.

Chapter 14

Mithradates, Pirates, and Rome

The pirates were an annoyance to Rome, though rather more threatening to those in the east. Antonius' campaign in 102–100 had a beneficial effect for a time, but it would take more than a single cruise along the Kilikian coast to suppress the problem. The collapse of the Seleukid government in Syria and Kilikia in the period after 128 further fuelled the general piracy problem, and from 89 a new war in Asia and Greece pumped up the supply still more.

During that same period (say, 110–90 BC) the king of Pontos, Mithradates VI, was restlessly attempting to expand his ancestral kingdom. Most of our information on this relates to the various attempts he made to take over the neighbouring kingdoms of Asia Minor – Bithynia, Kappadokia, Paphlagonia – but he was also active in the Black Sea area. One of his successful projects was to send an army and commanders to beat back the barbarian attacks on the Bosporan kingdom of the Crimea, and then to oversee that kingdom's expansion.

Mithradates therefore had a fleet. His own ancestral kingdom was spread along a thousand kilometres of the coast of the Black Sea, and included several Greek cities, such as Sinope and Amastris. The timber and iron mines of the inland mountains provided raw materials for ships. When he went to war with Rome he could produce a considerable fleet. It would seem to have been one of the projects he set going well before that war, and the conquest of the Crimea is one item of evidence for it. A speech attributed to Mithradates and placed before the Roman war contained the claim that he had 300 *cataphracts* – decked ships, usually of the trireme size at least, though this, judging by the actual numbers quoted in the wars, is a great exaggeration.[1]

When the war began, in 89, Mithradates swiftly conquered most of Asia Minor, including the wealthy cities of the Ionian coast, many of which had small squadrons of warships. Casual references in inscriptions and literary texts show that there were ships in Lykia, where there were enough ships among the several cities to make up a small fleet which campaigned against pirates soon after Antonius' expedition;[2] at the mouth of the Black Sea (either the Bosporos or the Hellespont) there was a small squadron when the war began whose sailors either fled or joined Mithradates, though it is not clear where the ships were from;[3] Byzantion, Ephesos, Chios, Herakleia-Pontike, Kos, and other cities are noted at various times as having warships, and most of them joined Mithradates, at least for a time. These places, of course, were originally all allies of Rome, and many had contributed to Antonius' anti-piracy campaign ten years before.

Elsewhere in the region there were ships at Athens,[4] and some under direct Roman control in Macedon, now a Roman province.[5] The Pamphylian cities were able to send several ships to sea;[6] probably many of the Aegean islands could send out one or two at least. But the main power in the Aegean Sea was still Rhodes. Many of the ships of those cities were available to Mithradates, though not all; he also organized the building of more for his own fleet.

The causes of the outbreak of war in 89 are complex, as is usual for most wars, but one which was mentioned more than once by Mithradates' envoy, Pelopidas, was that King Nikomedes of Bithynia had 'closed the sea', that is, that he was preventing the ships of Mithradates' subjects from sailing through the Straits, a serious matter for the merchants of the Bosporan kingdom in the Crimea.[7] One of the moves made by the Romans was to place two Roman commanders at the Bosporos and Byzantion to reinforce that blockade.[8]

Mithradates sent his own fleet into the Aegean as part of his campaign. They presumably chased off or seized the guard ships at the exit to the Black Sea, and collected ships from those who joined him, though not all were willing, and others were less than enthusiastic.[9] The major local enemy to his conquest was Rhodes, still the most important naval power in the region, if somewhat reduced by now. He brought his whole fleet south to attack the island. The Rhodian fleet came out to contest Mithradates' forces before they could land, but the king's fleet outnumbered that of the city, and threatened to outflank it, so the Rhodians retired to the harbour, apparently without coming to contact.[10] Later there were clashes between small Rhodian squadrons and isolated royal ships. One of these was a veritable battle, involving twenty-five royal ships against six Rhodian; the Rhodians, as usual, claimed to have won. Mithradates used a quinquereme as his flagship, and captured one Rhodian quinquereme; triremes and biremes are mentioned on both sides.[11] The monster ships of two centuries before were no longer being built; it was sea warfare as the fifth century Athenians had known it. It seems clear from the accounts of the fights that the Rhodian skills at sea had not diminished; Mithradates' siege failed.

He sent his general Archelaos to Greece with a 'great fleet'.[12] This indefinite phrase is misleading. Some of his warships had been left to watch Rhodes, and this squadron was big enough to prevent Rhodes taking much part in the fighting. Other ships were probably at the Straits or in the Black Sea, so this 'great fleet' was probably mainly transports, since its task was to move the army to Greece. At the same time, it is clear that it carried only a part of Mithradates' army.

Nor was the crossing completed without opposition. It was necessary to fight for Delos, where there was a large Italian population engaged in commerce. Since Mithradates had ordered the massacre of all Romans and Italians in his conquered territories in Asia, it is not surprising that they resisted.[13] And when Archelaos reached Greece, his naval forces were attacked by some ships from Macedon under the command the Q. Bruttius Sura, a legate of the Macedonian propraetor C. Sentius.[14] Bruttius later commanded a detachment of the Roman army in Macedon, which disputed control of Boiotia with the army commanded by Archelaos, fighting three drawn battles until he was sent back to Macedon. The whole Roman garrison of Macedon was only two legions, and Bruttius clearly had a smaller force than that. Similarly in the sea fight Bruttius' squadron sank 'one small ship and one *hemiolia*', which does not sound like a big fight, so large forces were not involved on either side.

This small result is possibly a sign of the generally small numbers of vessels employed by both sides. At Rhodes, the largest squadron was twenty-five ships, though at the beginning the Rhodian fleet was outnumbered. Of course Mithradates, once in the Aegean, had to spread his ships out, but Archelaos' great fleet conveying his forces to Greece was probably mainly transports, and even then he had to ferry the army over the sea in sections. In the fighting against Bruttius, who had only a fairly small Roman force, Archelaos cannot have had more than, say, 10,000 men at the most. This first Mithradatic army in Greece was only an advance force, sent to test the waters and secure local support.

So it would seem that nobody had many warships, and that the boast that Mithradates had 300 ships, made at the beginning of the war, was a gross exaggeration. The Rhodians were able to stand off Mithradates' attack, but they remained blockaded by a Mithradatic squadron. At Delos the first fighting was between an Athenian squadron and a Roman naval force under L. Orbius, probably a praetor.[15] This was not 'Roman' in the sense of being composed of ships from Rome or even Italy, but cannot have been other than a scratch force of allied vessels, perhaps including converted civilian ships gathered from the islands, unless they were Delian ships. When the Mithradatic 'great fleet' arrived, this 'Roman' fleet vanished.

One of the reasons Mithradates had launched his invasion of Roman territory when he did was that Italy had collapsed into a civil war. This was obviously an ideal moment, since the civil war had to be finished before he could be attacked. He had defeated the small Roman force in Asia, and for a time ignored the main Roman army in the region, the two legions in Macedonia. The Roman civil war lasted until 87, and then the Senate, dominated by L. Cornelius Sulla, the victor in the war, sent him to Greece with an army of five legions.

But when Sulla reached Greece, Italy behind him fell back into the control of his political opponents. He was thus unlikely to be able to receive any reinforcements. These opponents then sent another army to replace him, under the suffect consul L. Valerius Flaccus, so Mithradates found himself facing two Roman armies, hostile to each other, though Sulla ignored this rival army and fought on in Europe. Flaccus got his army through Macedonia to the Straits and across into Asia, though he had to run the gauntlet of a Mithradatic squadron operating off Brundisium, presumably from a base in Greece.[16] Sulla had found that a lack of ships hampered him during the sieges of Athens and Peiraios. He sent L. Licinius Lucullus to tour the eastern Mediterranean to collect a fleet from the Roman allies.[17] (Flaccus had clearly been able to find ships he could use at the Hellespont, though there is no sign that he had any warships; it would seem, therefore, that Mithradates neglected to leave a squadron to guard the Straits.) After defeating Mithradates' armies, first that under Archelaos, then another sent over from Asia under Dorylaos, Sulla began to build his own ships.[18]

At this point Lucullus returned. He had had to sail clandestinely, but during the winter of 86/85 he visited Rhodes, Pamphylia, Phoenicia, Cyprus, Egypt, and Cyrenaica, all of which apparently contributed ships to his force. This all took time, and the ships would clearly not be able to sail to Greece during the winter. It is a testimony to Lucullus' persuasiveness and authority, therefore, that he reached Sulla in Boiotia in the late spring.

On the other hand Lucullus had not gathered a fleet strong enough to challenge Mithradates' command of the Aegean. Mithradates was beaten by Flaccus' army, now commanded by C. Flavius Fimbria, and when Lucullus arrived Mithradates was under a land siege by Fimbria at Pitane. The king sent out a call for his ships to gather; Fimbria demanded that Lucullus blockade the port.[19] Lucullus refused to help and Mithradates escaped.

It is usual to assume that Lucullus' refusal was a result of the Roman internal dispute, for neither commander recognized the other's authority as legitimate. That may well have been part of Lucullus' calculations, but the more immediate problem for him was that Mithradates' fleet was stronger than his. If he put his fleet in a position to blockade Pitane, he would also be in a position to be attacked by a superior force. His primary mission was to join Sulla, and all his work of gathering ships during the winter would go for nothing if he allowed his fleet to be sunk or captured as soon as he arrived in Greek waters.

By this time Archelaos, who was in Chalkis, and with a good line of maritime communications to Mithradates in Asia, opened negotiations with Sulla. Both sides came to the talks with considerable assets, military and political, but were both under a variety of pressures. One of these for Sulla was his lack of ships. Lucullus' fleet was very miscellaneous, and unlikely to be willing to leave Greek waters, while the sailors would no doubt be happy to go home at the earliest opportunity. It was not a strong enough force to challenge Mithradates' own fleet, and Archelaos had no problem in communicating with Mithradates in

Asia. Sulla sent Lucullus to Abydos to seize control of the Straits, but Mithradates' ships would clearly be able to transit the Straits if they wished, and, once in the Black Sea they could prolong the campaign indefinitely.[20] Sulla's terms of peace were therefore lenient: Mithradates must withdraw to his own kingdom, give up his conquests, and surrender to Sulla that part of his fleet under Archelaos' command, stated to be seventy ships.[21]

Mithradates finally agreed to the terms, after meeting Sulla at Dardanos on the Hellespont. He was attended there by a large army, and 200 cataphracts.[22] By this time Archelaos had left Europe, so this fleet probably represents Mithradates' full naval force. Sulla, of course, still had only the ships Lucullus had collected. After some misgivings Mithradates accepted Sulla's terms, handed over the seventy ships, and returned to his kingdom.[23] When Sulla returned to Italy, therefore, he did so in a convoy said to be of 1,600 ships, of which the seventy Pontic warships were no doubt part. He landed at either Brundisium or Tarentum (or both). His enemy L. Cornelius Cinna had gathered an army at Ancona, where he was also collecting ships; he was stoned to death by his own troops before he could embark. Sulla won the next bout of civil war, and made himself dictator. The ships he had taken from Mithradates, now Rome's ships, constituted the first Roman fleet for almost seventy years.

Mithradates, if he had 200 ships at Dardanos, retained the larger part in the peace terms. He retired to Pontos and had to spend some time reasserting his authority in Colchis and in the Bosporan kingdom; a fleet was obviously useful in these endeavours. The removal of his fleet from Aegean waters, west to Italy and north to the Black Sea, seems to have been decisive for a sudden expansion of piratical activities. Lucullus' fleet was either taken to Italy or was disbanded and sent home, probably the latter. The ships Mithradates had collected from the Asian cities were either kept by him (or by Sulla) or, given the destitution of much of Asia after 85, laid up.

Many of the cities had rebelled against Mithradates at the end, and many had suffered for it. When he arrived to sort out the situation, Sulla imposed a massive collective fine of 20,000 talents, in addition to which all of the armies had indulged in looting. Piracy was then fuelled by the complete absence of naval opposition, by poverty, and by opportunity. Even as Sulla was travelling back to Greece from Asia, widespread piracy was developing; he is said not to have cared, since the victims had probably fought against him.[24]

The ancient historians generally link Mithradates with the pirates, implying, or stating, that he was largely responsible for the sudden growth in piracy which evidently took place in the period after about 90 BC. No doubt Mithradates did hire pirates to fight for him, both as soldiers and as sailors, but once hired, they were not pirates but mercenaries. We know the name of one pirate captain, Seleukos, who saved him from shipwreck in a storm; in the story Seleukos was not trusted by Mithradates' councillors, but he was by the king; that is, he was still regarded as a pirate by the councillors, even though employed.[25] If such men remained pirates they were neither under Mithradates' control nor subject to his orders. He certainly needed a large number of men, especially sailors, for his maritime needs were great. It was also a very useful insult for his enemies to throw at him, both to persuade Greeks to reject him, and to persuade Romans to fight him. But once peace was made, he cannot be held responsible for the actions of men whom he had employed and who, after being dismissed, reverted to being pirates.

Further, it is likely that the notices describing pirate activities which have survived are both distorted and exaggerated. As an example there is a notice in Appian's account of Sulla's war on Mithradates to the effect that a party of pirates raided a temple on Samothrace, looted it of a thousand talents' worth of treasures, and that this happened while Sulla was on the island. But there is no other indication of Sulla visiting Samothrace. When he left Asia he went directly to Greece, from Ephesos to Athens, not a journey which would take in a visit to Samothrace. So this story looks like an invention, or a misunderstanding of Appian's source. It may well be that Samothrace was raided, but not while Sulla was there, and the loot worth a thousand talents looks to be a surprisingly large and round number.[26]

However, while the details may be suspicious in some cases, there is no doubting the widespread activities of the pirates. The link with Mithradates is in all likelihood no more than propaganda, together with the fact that during his wars naval forces in the eastern Mediterranean were generally preoccupied with other duties, and pirates could therefore become active with less fear of opposition than in peacetime. As an example there is Lucullus' experience during his long attempt to gather a fleet for Sulla's use. He was attacked by pirates while sailing west from Egypt, and at the city of Berenike in Cyrenaica he found that it had been twice attacked by pirates in the previous two decades. He spent some time (presumably while waiting for his ships to appear) regulating the city, though he did not rebuild the ruined walls. On his way to Greece with his fleet he deliberately avoided doing anything about piracy. He had other priorities.[27]

Once Sulla returned to Italy with his confiscated ships, and Mithradates withdrew to the Black Sea, only the local forces of the eastern states were available for anti-pirate duties; that is, the local situation reverted superficially to the situation as it had been before the war. But in the meantime those local forces had been reduced, by ships sunk, confiscated, stolen, or whatever other reason, while the attractions of the pirate life had clearly increased. Appian states that Iasos, Samos, and Klazomenai, as well as the Samothracian temple, were 'captured' by pirates.[28] Such raids could only have been conducted by large bands of pirates, and it is this development which seems to be the main reason for the increased threat felt by the settled world. Captain Seleukos, who saved

Mithradates, may have led one of these bands; another was led by a man called Isidoros.

The issue was complicated by the presumed connection of the pirates with Mithradates, not necessarily because the king was directly involved with them, though no doubt he was always pleased to see his enemies distracted by pirates' activities, but because, as perceived at Rome, he himself was the major problem. As a result, whenever a decision at Rome was taken to attend to the issue of piracy, it tended to involve Mithradates as well.

Sulla, who was familiar with the east if anyone at Rome was, seized power on his return to Italy, but died in 78. There is no indication that he bothered about piracy during his brief rule, but the Senate did attend to it in 78. The consul for 79, P. Servilius Vatia, was given Kilikia as his proconsular command for 78 and after. He had been Sulla's choice as consul, and this may have been one of the dictator's last decisions. Vatia campaigned along the coast of Lykia and Pamphylia, punishing those cities he felt were involved with pirates, but his actions were inconsistent, and were eventually shown to be directed more against Mithradates than the pirates.

He found that two cities, Olympos and Phaselis, had fallen under the control of a local chieftain, Zeniketos, and that Phaselis and Attaleia were providing markets at which pirates could sell their loot. He eliminated Zeniketos, set up the two cities in independence once more, and punished Phaselis and Attaleia by depriving them of some land.[29] The next two cities on the route east through Pamphylia were Perge and Aspendos; both had been visited two years before by C. Verres, acting for the governor of Asia, Cn. Cornelius Dolabella. Later Cicero accused Verres of stealing various things from the cities, but he actually went there as Dolabella's legate, and the only reason for him to do so was to ensure that the cities were free of pirates and not involved with them.[30] So between them Verres and Vatia had cleared most of Pamphylia of the pirates and their influence.

Vatia did not go to either Side or Korakesion, though the former was notorious for its hospitality to pirates in its market, and the latter is specifically described as a pirate town.[31] Nor did he even approach Kilikia Tracheia, the main source of pirates. Instead he turned inland after visiting Aspendos and spent most of his proconsulate establishing Roman control in southern Lykaonia. This was therefore a move in the wider diplomatic and military campaign against Mithradates, in preparation for the next war. Vatia's command against Kilikia was therefore not primarily anti-pirate, but mainly anti-Mithradates.

This essentially neglectful attitude of Rome may well have encouraged the pirates, though they scarcely needed it, but Rome's actions worried Mithradates more. One of his responses was to contact Q. Sertorius, a Sullan enemy who had maintained himself against senatorial authority in Spain for several years.

Mithradates and Sertorius exchanged envoys. Sertorius sent M. Varius, who attempted to take over as governor of Rome's Asian provinces; their enemies claimed that Sertorius was actually handing over the Roman provinces to Mithradates.

Both sides were therefore gearing up for a new war in the east. When the first war with Mithradates began in 89 there had been effectively no Roman forces in Asia, and just two legions in Macedonia, but by the time Vatia had finished his work in Asia there were two legions in the Asian province and two in the Kilikian command, while Roman political control had also been extended over a large area of southern Anatolia. By 74 both sides decided war was imminent, and both began to build up their naval strength in the region as well.

Mithradates began, once again, with a substantial fleet, partly retained from the last war, and partly of ships built then. At Rome, however, the ships at the city's disposal are not stated at any time. It is necessary to detect Rome's naval strength by looking at what was happening. It becomes evident that a steady expansion of Rome's navy was taking place.

The Roman fleet was based on the ships Sulla brought back from the east, with others either newly built or otherwise acquired. Vatia in Pamphylia may have had some ships, though his campaign was essentially by land, and he may merely have used locally conscripted ships in the same way as Antonius had a quarter of a century before. Sertorius hired some Kilikian pirates when he was driven into Mauretania and used them in an assault on Ibiza, from which he was expelled by the local governor, who arrived with a fleet of his own.[32] During his campaigns in Spain Sertorius maintained a small fleet which helped delay the conquest of that land by the senatorial generals.[33] What sort of ships these were is not made clear, and most of them could well have been simply transports, since the various activities described are landings and voyages rather than battles. (The presence of Kilikian pirates in the westernmost Mediterranean, however, is a sign of their ubiquity.)

The Senate, facing Sertorius in Spain and Mithradates in the east as well as the pirates, also faced a crisis in the supply of food to Rome itself in 75–73. In 74 the Senate responded to all these accumulating problems. The senatorial commanders in Spain, Cn. Pompeius Magnus and Metellus Pius, were helped with supplies and money; the consuls of 75, Lucullus and M. Aurelius Cotta, were assigned to the war with Mithradates for their proconsulates, and it duly broke out next year; and one of the praetors of 74, M. Antonius, was assigned a sea command.

Antonius, the son of the Kilikian campaigner of 102–100, had a vague command which changed as circumstances changed. He first had to gather his forces, which he apparently did first of all in Sicily, but also perhaps elsewhere. Since his command was specifically at sea, he clearly needed a fleet. His activity in Sicily, which is never spelled out, would seem to have been directed at gathering ships and crews from the naval allies there.[34] No doubt he also collected others from the naval allies in southern Italy. But he needed a more substantial force than could be acquired in this way, and it may be that he based his locally recruited fleet on a squadron of Roman ships.

Antonius' first task appears to have been to secure the sea route to Spain, for one of his recorded exploits involved a landing in Liguria to suppress a pirate community. He had a fleet with him, but what it consisted of and how big it was is never stated. After Liguria he moved on to Spain, but his activities there are not known.[35] These activities, and others, seem to have occupied him during 74 and into 73, and were obviously connected with the new determination of the Senate to finish off the Spanish war. The death in 73 of Sertorius, who was being defeated anyway, allowed the rebels in Spain to be suppressed, so releasing both soldiers and ships for other duties.

Antonius was now given the task of attacking Crete. This was, of course, an old source of piracy, and no doubt the improved opportunities of recent years had been seized on by these old practitioners. Antonius spent time gathering resources in Greece – much to the discomfort of the Greeks – then attacked Crete and was beaten.[36]

The Cretans were said to be sympathetic towards Mithradates, and it is clearly relevant that Antonius was directed to make his attack while the Third Mithradatic War was on. His defeat scarcely mattered, for by the time he invaded Crete the new war with Mithradates had been on for two years. But his campaigns suggest that, while Roman sea power had developed substantially during the 70s, it was still by no means overwhelming. His Cretan adventure had been delayed by his need to gather more forces in Greece, so it would seem that whatever forces he had in the western Mediterranean during 74–73 had been taken from him, and he had to start again. Once more the Mithradatic war had taken precedence.

Mithradates is usually blamed for starting the new war, but Roman preparations – Vatia's campaign, for instance – were clearly designed to establish a position from which the king could be attacked. He certainly moved first, by invading Bithynia, which had just been made a Roman province. This time the Romans were more or less prepared. Both consuls were in Asia, Lucullus as governor of Asia and Kilikia, with five legions. M. Aurelius Cotta, Lucullus' consular colleague of the year before, was governor of Bithynia, and he was allocated a substantial naval force.[37]

Mithradates used his naval force to gain early command of the Straits. He already had some ships in the Mediterranean, sent to contact Sertorius, but they arrived at Spain only after Sertorius was dead, and the Romans were expecting them to return to the Straits.[38] Another squadron was sent to Crete,[39] and perhaps reinforced the resistance of the islanders to Antonius. At the very least the presence of Pontic ships in Cretan waters will have been one of the reasons Antonius took so long to get to the island.

Before the war began, therefore, Mithradates had sent these squadrons through the Straits. One of the preoccupations of the Roman commanders was the realization that their own fleet, which was brought to Chalkedon on the Bosporos, now lay between two Pontic naval forces. Mithradates brought his main force west by sea, captured Herakleia Pontike, and drove Cotta into Chalkedon. Mithradates' fleet stormed the city's harbour, burning four ships and capturing sixty. This was presumably the immediately available Roman naval force. It was, since it had been sent to Chalkedon, clearly intended to block egress from the Black Sea by controlling the Bosporos. Cotta's naval commander, P. Rutilus Nudus, commanded in the land battle, presumably using the soldiers from the fleet. He was defeated with the loss of 3,000 men, so his fleet was effectively beaten before Mithradates' ships arrived.[40] This would explain the ease with which the Roman ships were taken.

These vessels were, as usual, only 'Roman' in the sense that they were under the command of Roman officials. The fleet at Chalkedon was one gathered from the Asian, and perhaps the Greek, naval allies; one group of ten was from nearby Kyzikos, and it seems likely that most came from cities around the Propontis and the Straits. But having lost them, the Romans perforce had to fight on land. Mithradates seized control of many of the cities in the area. Lampsakos became his naval base, which gave him control of the Hellespont and the passage to the Aegean.

Mithradates besieged Kyzikos, which resisted unexpectedly well, and Lucullus in turn besieged the besiegers. Mithradates' huge army could not be properly supplied when the siege dragged on into the winter, even though he had command of the sea. He finally retired to Nikomedeia, where he intended to winter, though he lost part of his army in the meantime. The Romans collected another fleet from cities around the Aegean, to be commanded by Lucullus' naval legate, C. Valerius Triarius. Inscriptions record ships coming under his command from Miletos and Smyrna.[41]

Lucullus besieged Lampsakos and Mithradates rescued his besieged forces there by sending part of his fleet, fifty ships, to collect them.[42] As a result, the Hellespont was available to the Roman ships and Lucullus appointed it to be the rendezvous for his new fleet. A squadron of thirteen of Mithradates' ships was spotted nearby. Lucullus took several ships out to capture them,[43] and swiftly followed this up by catching part of the Pontic fleet at Lemnos (evidently a rendezvous) where most of the ships had been beached. The fight was thus on land, and the Roman troops were tougher and more skilled, and perhaps more numerous, then Mithradates' soldiers.[44] The fleet, perhaps eighty or ninety ships, which Mithradates had sent to cruise in the Aegean and about Crete, presumably with the intention of cutting Lucullus' communications with Rome and influencing local opinion in his own favour, now came north. It was tackled by the new Roman fleet, commanded by Triarius, in a battle near Tenedos, and the Roman fleet was again victorious.[45]

Mithridates had lost perhaps 200 ships in these battles, though he is said to have begun the war with 400. His fleet also suffered on at least two occasions from storms. One fleet, possibly that evacuating people from Lampsakos, fifty ships strong, was scattered by a storm in the Propontis, and many of the ships were sunk. They were on the way to Nikomedeia, Mithradates' new headquarters.[46]

There had also been other fighting. The Cretan force was about eighty strong at Tenedos, but had already lost ships in another storm, and in some 'small battles'.[47] So, while it was on campaign in the Aegean and about Crete, there had been Roman warships numerous enough at sea to tackle it, or perhaps detached parts of it. (And it must have been over a hundred strong to begin with.) One would suppose that M. Antonius in Greek waters, and perhaps in the west, was largely responsible. He was certainly awarded a triumph at the end of his proconsulate, despite his original defeat at Crete.

Mithradates retained a substantial fleet of over a hundred ships even after these defeats. He retired further towards his own kingdom. It was while sailing from the Bosporos towards Pontos that he was rescued by the pirate captain Seleukos, but much of his fleet was wrecked by that storm. The Romans planned a new offensive. Cotta by land besieged Herakleia Pontike; Lucullus marched the main army into Kappadokia to invade Pontos from the south.

The Roman fleet now controlled the southern coast of the Black Sea as well as the Aegean, and this must be the main reason for Mithradates to take the difficult route from Armenia north to his Bosporan kingdom along the eastern coast of the sea, marching by land, when he made yet another retreat in 66. He had shown himself adept at retreating into inaccessible regions, and no attempts were made to reach him in his last kingdom, though Roman naval command of the Black Sea remained necessary until he died.

The preoccupation of the Romans with Mithradates had once more allowed pirates to roam widely. Once more one reason was that the local warships had been conscripted into the Roman war fleet, leaving many coasts and islands without maritime protection. Antonius had some effect in the west, but in the east no suppression measures were effective. Plutarch provides a list of places in and around the Aegean sacked and raided by pirates in these years, eleven places in all, and others can be added. Delos was sacked, for the second time, by a pirate leader called Athenodoros, after which Triarius built a protective wall.[48]

Roman attention shifted back to the pirates, partly because of the sack of Delos, but also because Rome's own food supplies were again threatened.[49] Without an active, major enemy, the activities of the pirates became more noticeable. After considerable intrigue at Rome a command against them was given to Cn. Pompeius. It was an enlarged version of the successive commands given to Antonius, and included authority to levy a fleet and an army, employ legates, and to exercise power up to fifty miles inland from the coast. All this shows a full understanding of what was needed to defeat the pirates: a fleet to drive them ashore, and an army to capture their bases.[50]

The main account we have of Pompeius' pirate war is by Appian, as part of his history of the wars against Mithradates, even though it bears all the hallmarks of the exaggerations to be expected of Roman party political propaganda. He speaks of pirate numbers as 'tens of thousands', of their bases in forts, on peaks, and desert islands, 'everywhere'. They were originally 'of almost all the eastern nations'. He emphasizes the failures of the various Roman attempts at suppression, the defeats inflicted 'on Roman generals in naval engagements' (which must mean Antonius at Crete), and 'among others the praetor of Sicily on the Sicilian coast itself'. Murena and Vatia 'accomplished nothing worth mention'; in the case of the latter this is simply not true, and Murena did not campaign against pirates.[51]

This reflects the propaganda put out as a means of persuading the Senate to entrust the command against the pirates to Pompeius. Yet the pirates also co-operated in bringing about their own destruction. Athenodoros' sack of Delos, a holy island, shocked large numbers of religious Romans, as well as hurting many Roman companies who traded through the island. More immediately the supply of grain to Rome was again being interrupted, and pirate raids were made on the Italian coasts; a 'Roman fleet' was even attacked near Ostia; 'some women of noble families' were kidnapped (including the daughter of M. Antonius, ironically), and 'two praetors with their very insignia of office' were captured.[52] All this was even more convincing than the exaggerations of pro-Pompeius propaganda. At least one member of the Senate, C. Julius Caesar, could testify to their methods and ubiquity, having himself been captured and ransomed. He collected a force after his release and cleared out that particular nest, but such minor actions had little overall effect.[53]

The fleet Pompeius had to use was originally set at 200 ships, and later at 500, though the highest number he is said to have employed was 270.[54] He followed Antonius' lead in clearing the western basin of the Mediterranean first, accomplishing this in forty days, supposedly.[55] His priority was clearly to restore the Roman grain supply, and so he concentrated on Sicily, Africa, and Sardinia. The speed of this campaign implies that it was somewhat superficial, and that the earlier work of Antonius had been effective. No doubt also, to campaign in the east, Pompeius would need to move fast, before his political enemies could impose restrictions on him. The campaign in the east took longer, but he was assisted by a simultaneous campaign of conquest in Crete by Q. Caecilius Metellus, though Metellus refused to accept that his own authority was overridden by Pompeius'. By assigning specific areas to his legates the pirate bases could be more or less simultaneously suppressed. Kilikia Pompeius reserved for himself.

Whereas in the west only the fleet and its marines were needed, in the east Pompeius required legionary troops as well. He was authorized to recruit 120,000 soldiers, but probably needed less than a third of that. (Large armies need much feeding, and can break down under their own weight, as Mithradates at Kyzikos had discovered.) The pirates resisted. The battle against their fleet took place off Korakesion; at least three places on land had to be captured, including Korakesion itself. As important as these victories was his capture of 71 ships; 306 were surrendered, according to Appian. Of these, according to Plutarch, 90 were warships armed with rams; according to Strabo, 1,300 ships and boats were burned. None of these figures may be accurate, but the overall impression is clear: the naval power of the pirates was broken, and by destroying ships, and wood, gear, and sails, its recovery was prevented. Those who surrendered were given the option of being settled in cities in Smooth Kilikia and Greece, and most of them accepted. This was not the end of piracy in the Roman Mediterranean, but it was never again to be the menace it had seemed. The speed and ease with which Pompeius succeeded is a testimony to his ability as an organizer; it is also, no doubt, a testimony to the exaggerations of the earlier anti-pirate propaganda.

The war with Mithradates was still unfinished, and Pompeius was given this task also; he took the opportunity to acquire large areas in Asia Minor (including Pontos) and Syria as new Roman provinces. These acquisitions pushed Mithradates away to his Bosporan kingdom (he died in 63), and established a firm Roman control of all the coastlines of Asia Minor and Syria; this had been an area of constant warfare in the past generation, conditions which encouraged piracy, so these annexations were in fact the culmination of the anti-piracy campaign.

One of the effects of all this eastern warfare had been to re-establish Rome as a naval power. The numbers of ships Pompeius was authorized to use were much greater than those available from the naval allies, and anyway many of their ships were still in use in the war against Mithradates. Only by new construction over the past twenty years in response to the needs of earlier campaigners could Pompeius have gathered so many ships so quickly. At the end of his eastern campaign, therefore, Rome had a navy again, of a substantial size, for the first time in a century.

Chapter 15

Roman Civil Wars

One of the elements in this tale of sea power which has emerged repeatedly is that it was relatively easy to develop a fleet of vessels of war. This had been done by Mithradates, just as it had by Rome and Antigonos I and Antiochos III, and even by Philip V. It was the relative ease of building such warships which was the cause of the fear developed in his enemies by the plans of Demetrios Poliorketes, in Rome by Carthage, and was at the root of the threat of Mithradates' power. He had twice built up a major fleet which dominated the seas from the heel of Italy to the coast of Anatolia.

That this was done so often in the Hellenistic period is one of the major elements of power in that time. It presupposes a supply of materials, notably wood, rope, and canvas, the presence of a skilled woodworking shipbuilding population, and the ability of governments to recruit crews – oarsmen, marines, sailors, officers – with ease. The locations where this was achieved were fairly concentrated – Macedon, Phoenicia, South Italy, Pontos – the areas where the wood was available. This is reasonable, since it was wood in bulk which was the most difficult material to move. The presence of suitable forests therefore was the first essential in developing a state's sea power.

This was also one of the bases for the superficially strange Roman practice of allowing its fleets to rot away. Only repeated wars kept a fleet in existence, as between the first fleet, constructed about 260, which was kept going with maintenance and replacements until the 170s. After that there is no sign of one, and the city relied on conscripting ships from its allies and subjects, who, by contrast, continued to maintain small forces. Of course, since this turned out to be sufficient – in the 140s, the 130s, and again largely against Mithradates – it could be argued that Rome had no need of a fleet of its own. On the other hand, had it maintained one, it seems doubtful that either the pirates or Mithradates would have posed such a challenge in the first century BC.

The fleet acquired from Mithradates at the peace of Dardanos in 87 – seventy ships – may have been used in the later wars, but by the time the next great crisis for the Roman state developed – the civil wars (49 to 30) – those ships will have gone out of use for the most part. So in the civil wars another collection of allied ships was made, and this is another example of the speed with which a large fleet could be built.

In the aftermath of the Mithradatic wars and the suppression of the pirates, for some time Roman aggression turned outwards; Pompeius seized Syria and, soon after, Caesar conquered Gaul. In the process Caesar had to build a fleet on the Loire, first to combat the seamen of the Veneti of Armorica, then to transport his army to Britain, but these ships were apparently abandoned later. When he came back to Italy in 49, the civil wars began, lasting on and off for almost twenty years.

There is no sign of the existence of a Roman fleet during the first crisis of this new civil war. Caesar's conquest of Italy pushed Pompeius back to Brundisium, whence his soldiers were shipped across the Adriatic to Dyrrhachium; Caesar besieged him at Brundisium, attempting to block the harbour with an artificial breakwater; Pompeius and his forces escaped. Neither man seems to have disposed of any warships, though a reference to some 'swift ships' in Caesar's account of the escape of the last Pompeian troops might imply a galley force there.[1]

When Caesar headed for Spain later in 48, Massalia refused him entry. It had accepted L. Domitius Ahenobarbus, who was the Senatorial appointee to succeed Caesar in Gaul, when he arrived with seven ships. These had been requisitioned at Cosa and the Iguvian Islands, and manned with Ahenobarbus' own slaves and tenants. It is clear they were warships.[2] Caesar left three legions to form the siege and went on to Spain. He had twelve new warships built at Arelate (in thirty days) and put D. Junius Brutus in naval command;[3] Brutus had experience with sea warfare against the Veneti in Gaul. The Massaliots added their own ships to Ahenobarbus' seven and came out to fight.

The ships involved on both sides were mainly *cataphracts*. Brutus' fleet of twelve ships were heavy vessels, probably quinqueremes; the Massaliot ships were similar in size, but not in strength, so they were of a lighter build but still quinqueremes. The seven ships brought by Ahenobarbus were certainly warships, but are likely to be either triremes or, more likely, liburnians (that is, much like penteconters or *lemboi*).

In the first battle, the heavier and better-manned, though outnumbered, Caesarean ships defeated the Massaliot squadron, capturing six ships and sinking three.[4] This now made Brutus numerically the stronger, and so the Massaliots remained in port. From Greece, Pompeius sent a squadron of sixteen ships under L. Nasidius, to which he added another by a raid on Messana as he passed the Strait. The Massaliots brought back into service some old ships, so with Nasidius' squadron they now once again had the larger fleet.[5]

A second battle followed. This was essentially between Brutus' fleet and the Massaliots, for it turned out that Nasidius' ships were too small and light, and perhaps too lightly manned, to participate. He took his ships out of the line of battle and headed for Spain. Again the heavier Caesarean ships prevailed and sank five of the Massaliots, who also lost four ships captured. Nasidius also took one away with him to Spain.[6]

The tactics employed mirrored those of two centuries earlier in the great battles of the first war with Carthage: the Greeks of Massalia attempted to ram, and the Caesareans used the weight of their heavier ships and their more numerous soldiers to board. As they had against Carthage and Antiochos III, the more brutal tactics prevailed over the more elegant, though in one case Brutus' flagship was about to be attacked by two smaller ships from both sides, but a sudden acceleration left the ships colliding with each other, so there was an element of farce as well.

The sources of these ships are telling. Ahenobarbus found his seven ships at small Etrurian ports, but they had to be 'improved' to be able to stand in the line; Massalia had a squadron of eleven ships available, and a group of older ships which could be fitted out. Nasidius brought sixteen ships from Greece, and found another at Messana, which he took from the dockyard, where it was either being built or repaired. These are examples of the local ships and fleets on which Rome had relied earlier. Probably Nasidius' ships had the same origin. But Caesar, in control of all Italy, including Rome and Ostia, and the Campanian and South Italian ports, had to build his own squadron.

There is also a description of the battle at Massalia in Lucan's poem *The Civil War*, though the poet's verbal imagination makes it somewhat florid in tone. He claims that Brutus' ships were quinqueremes – which is likely – and that the flagship was a six. This is nowhere even hinted at in the contemporary account by Caesar himself. On the other hand, the flagships of the imperial fleets were sixes, as were those of some of the Roman commanders in the first war with Carthage. So it would seem that Lucan fell into the trap of assuming that the flagship would be a six. Given this presumption it is difficult to accept the details of the battle which he gives are any more accurate.[7]

It is also noteworthy that there were many merchant ships plying the seas. Pompeius had no difficulty in ferrying his large army across the Adriatic, and other men got across to Africa and Sicily and Greece without difficulty. In Massalia, before being shut in, Ahenobarbus sent out his seven ships to bring in merchant ships in the local seas, and several were quickly taken.

Caesar collected ships here and there, though without much system, and when he crossed the Adriatic he congratulated his soldiers for conquering a sea without a warship.[8] Appian says that Pompeius captured forty of Caesar's ships, though they seem to have been transports, not warships.[9] When a Caesarean fleet under Scribonius Curio sailed from Sicily to Africa, it had an escort of twelve warships, which were presumably from the Sicilian cities;[10] in Africa, Curio's opponent Attius launched ten warships which had been laid up after the pirate war and preserved; they were at least fifteen years old, perhaps twenty.[11] In Spain a dozen warships were built at Gades and others at Hispalis, and these were taken over by Caesar when he won the first part of the war there.[12] Nasidius took his fleet from Massalia to Spain, and seems to have gone on from there to Africa.[13] In the east Pompeius gathered ships from ports from Pontos to Cyrenaica; the areas especially noted were Asia, the Kyklades, Athens, Pontos and Bithynia, Syria, Kilikia and Phoenicia, and Egypt. He collected 500 vessels, plus some liburnians, according to Caesar,[14] or 600 according to Appian, which number presumably included the liburnians.[15]

Pompeius brought his fleet to the Adriatic, organized it into five squadrons, and distributed them along the eastern Adriatic shores.[16] The Adriatic became the front line in the war for a time. Caesar, despite the overwhelming numbers of Pompeius' fleet, got his army across in transport ships.[17] He was then cut off from reinforcements and supplies, though he was able to block landings by the enemy ships, forcing them to use only the ports they controlled.

The Pompeian ships attempted to isolate and blockade Caesar's forces, but found it very difficult. In the spring of 48 M. Antonius crossed the sea with a fair wind with more reinforcements in transports. The Rhodian squadron of Pompeius' fleet, commanded by C. Coponius, came out to intercept, but was outsailed and then caught in a storm which wrecked most of the Rhodian ships.[18] Antonius' force arrived safely, and his junction with Caesar's forces changed the situation in the Balkans.

The Pompeian blockade was inevitably productive of a series of small naval actions. Caesar evidently was able to gather some ships, partly by captures, partly by levying them from ports he controlled. He had a small squadron of twelve ships at Orikon, which was raided by Pompeius' Egyptian fleet: four were captured, the rest burned.[19] Two Pompeian commanders attempted to seize control of Brundisium harbour by raids across the sea, but were unable to supply themselves and had to retire, with the loss by capture of a quadrireme.[20] A raid on Antonius' landing place resulted in the burning of thirty small ships.[21] A large squadron commanded by C. Cassius, made up of ships from Syria and Kilikia, sailed across to Sicily. A Caesarean squadron of thirty-five ships at Messana was burned, but a second squadron at Vibo resisted, losing five ships burned, but capturing two quinqueremes and two triremes out of Cassius' squadron.[22] These Caesarean ships were vessels which had been collected from here and there, but it looks as though shipbuilding was also going on.

Caesar abandoned the siege of Pompeius at Dyrrhachion, marched into Thessaly, and when Pompeius came up, defeated him in battle at Pharsalos. Pompeius escaped to Egypt and death. Caesar pursued him to the Hellespont. Here there is a story that his very presence frightened C. Cassius into surrender, though another version has it that it was Lucius, his brother.[23] In the story, which is doubted, Caesar is said to have acquired seventy ships. This may well be the only acceptable part of the tale, for he was now able to sail directly to Rhodes and then to Alexandria with a fleet of ships collected from several cities. Until then these ships were part of Pompeius' fleet, so they changed sides at some point, and the Hellespont is the first place where this seems likely to have occurred. In Egypt Caesar found himself blocked up at Alexandria. He had a small squadron of ten Rhodian and some Asian ships with him as escort,[24] and found he was facing an Alexandrian guard squadron of twenty-two ships, all *cataphracts*, plus the Egyptian force of fifty ships, quinqueremes and quadriremes, that had been sent off to join Pompeius and had since returned home.[25] In the fighting these ships were burned. Caesar sent for help from Rhodes, Syria, and Kilikia, and a fleet arrived nearby.[26] An attempt by Egyptian ships to prevent this new fleet from entering Alexandria harbour brought on a battle in which one quinquereme was sunk out of the Egyptian squadron, another was captured, and many of the soldiers on board the others killed.[27]

Caesar claims that, in various ways, the Ptolemaic government had lost 110 ships, and another force was now formed from all their remaining ships. This produced a fleet of twenty-seven quadriremes and quinqueremes, and an unknown number of smaller ships.[28] (The Ptolemaic fleet therefore had consisted of at least 137 quinqueremes and quadriremes, undoubtedly the largest fleet in the Mediterranean, until Caesar began destroying it.) Caesar's fleet included thirty-four ships, gathered from Rhodes, Pontos, Lykia, and Asia, ten quinqueremes and quadriremes, the rest smaller. The Egyptian attack failed with the loss of six ships.[29] A later attempt to cut off Caesar's supplies also seems to have failed. Caesar was in the end rescued by the arrival of reinforcements by land.

This Alexandrian War kept Caesar occupied and largely out of touch for the whole winter of 48/47. Meanwhile, his Roman opponents regained their balance. They still had considerable forces available and their fleet was still very large. It was concentrated at Corcyra when Pompeius was beaten, at a strength of 300 ships including those under Cassius, who had been raiding Sicily when he heard the news of Pompeius' defeat. The ships are stated to have been triremes. The commanders divided the fleet between them, probably taking a hundred ships each, plus the land forces nearby. Cassius was assigned to Pontos, Q. Caecilius Scipio and M. Porcius Cato went to Africa, which was already in the hands of one of their friends, and Cn. Pompeius went to Spain, where he quickly contrived to change its loyalties.[30]

This, of course, yielded the central strategic position, Italy, to Caesar, who exploited it intelligently. A preliminary campaign out of Alexandria quickly dealt with Pontos. It appears that Cassius had not followed through with the plan to use his part of the fleet to raise Pontos against Caesar, but Mithradates' son Pharnakes made an attempt to recover his ancestral kingdom anyway. Caesar rapidly stopped him. What happened to Cassius' share of the Pompeian fleet is not known precisely (though the seventy ships Caesar is said to have acquired at the Hellespont were probably part of it). It seems likely that the rest went to their home ports. Cassius himself was with Caesar in the Pontic campaign, so he had quickly switched sides as soon as Caesar had got out of Alexandria.

So the remainder of the civil war, for the moment, involved Caesarean invasions of Africa and (again) Spain. Of these only the first involved serious sea warfare, and it is a good example of the weaker force at sea defying its stronger enemy successfully.

The senatorial party in Africa had several fleets available. The incumbent governor Attius Varus had about twenty ships (ten captured from Curio, ten of his own); Scipio and Cato had brought about a hundred from Corcyra; M. Octavius had brought an unknown number after being defeated in the Adriatic; and some of Cn. Pompeius' ships went to Africa; the squadron of Nasidius from Massalia and Spain was also there, it seems. A fleet of something over 150 ships had been gathered; 'several fleets', according to one source.[31] Caesar neutralized this formidable naval force by simply crossing to Africa from Lilybaion in the winter. He did not give his captains a particular destination, probably because he was trusting to luck, and would simply land where he arrived, and he set off with an escort of no more than 'a few' warships, though he had others available.[32] With the ships which had stayed with him, he eventually landed at Hadrumetum, and in the following days the rest of his ships came in. By the time an enemy naval reaction came – the ships had to be launched and manned, having been laid up for the winter – Caesar had forty warships with him, divided between Thapsus and Hadrumetum. Attius Varus brought a fleet of fifty-five ships to Hadrumetum and attacked the smaller group; he was then himself attacked by Caesar with the bigger force and driven into Hadrumetum, but the ships were there beached. Both sides sank any enemy merchant ships they could find.[33]

For some unknown reason no more use was made of the Senatorials' considerable naval superiority. Caesar's warships never numbered more than the forty used against Varus' fleet, whereas they should have been able to muster at least three or four times that number of ships. It may be that they were short of manpower, for Varus used locally recruited (or impressed) Gaetulians as oarsmen in his fleet. When they were eventually defeated in battle, on the other hand, the several leaders were given their choice of the ships still at Utica to escape. Scipio was chased by a Caesarean squadron and committed suicide.[34] That squadron may be another clue. Its origin is not known, but it may well be that the narrow sea between Sicily and Africa was dominated by the considerable Caesarean fleet which he had left based at Lilybaion. The threat of its presence, combined with a shortage of naval personnel and the winter weather could have kept the Senatorial fleet in harbour.

Caesar enjoyed his conquests for only a little time. After his assassination in March 44 civil warfare resumed. The division between the Caesareans and the Senatorial party – now calling themselves Liberators – reopened. By 43 fighting resumed, and fleets were again being collected. Sextus Pompeius, younger brother of Gnaeus, had collected a following in Spain, and the Senate appointed him to 'command of the sea'. He collected ships at Massalia to add to those he brought from Spain (probably some of those his brother had taken west from Corcyra), and took over Sicily with its ships.[35]

In Syria the governor appointed by Caesar, P. Cornelius Dolabella, gathered a fleet in the old way, from the local naval sources: Rhodes, Lykia, Phoenicia, and Kilikia, and seized Laodikeia-ad-Mare as his base. Many of these ships must have been the same ones which had been in Pompeius' and then Caesar's fleets. The leaders of the assassins, M. Junius Brutus and C. Cassius, had decided to seize control of the eastern provinces. Cassius came to Syria and demanded ships from the same sources as Dolabella. Only the city of Sidon responded, and its fleet was beaten outside Laodikeia by Dolabella's fleet. Cassius repeated his demands and this time two Phoenician cities, Tyre and Arados, responded, while Rhodes declared neutrality and presumably withdrew the Rhodian ships from Dolabella's fleet.

Kleopatra of Egypt had recovered control of Cyprus as a gift from Caesar, and had clearly spent considerable resources in rebuilding the Ptolemaic fleet which had been largely destroyed in the fighting at Alexandria in 48/47. Part of that fleet was in Cyprus. Cassius contacted both Kleopatra at Alexandria, and her governor in Cyprus, Serapion. Presumably he knew that Kleopatra would refuse to support him, or prevaricate (she did the latter), but Serapion responded at once, sending, like the Tyrians and Aradians, 'all the ships he had'. With these Cassius eventually defeated Dolabella's fleet and then succeeded in taking the city.[36]

Neutrality in a civil war was not acceptable. Both the Rhodians and the Lykians had tried to deny Cassius, and now he turned on them. His argument was that their fleets could be used to threaten his communications, especially at the Hellespont, and that to leave them hostile, or even neutral, in the rear was too dangerous. Brutus moved against the Lykians, Cassius took his fleet to Myndos and prepared to invade Rhodes Island. He had eighty ships of the heavier variety, probably mainly quinqueremes; the Rhodians launched only thirty-three ships, certainly of the lighter type, perhaps mainly quadriremes and triremes. By taking the offensive the Rhodians gained initial surprise, but the heavier and more numerous Roman ships prevailed. It was once again a fight of lighter, nimble ships manned by experts against heavier, slower, and heavily manned ships; as usual Roman weight counted, though it was largely because of their greater numbers. The Rhodians lost five ships; the rest returned to the harbour. After a brief siege and another naval engagement, Cassius was let into the city by a party favourable to him.[37]

Kleopatra was rumoured to be sailing with another fleet direct from Alexandria to Greece in order to join Octavian and Antonius. A fleet of sixty ships under L. Staius Murcus was posted at Cape Tainaron on the southern end of the Peloponnese to intercept her. But she was ill and the fleet did not sail. Murcus was sent on to prevent Antonius crossing from Brundisium; Cassius took his victorious fleet to the Hellespont to supervise the crossing to Europe of his army.[38] In his time at Rhodes and in Asia he cleared the whole area of treasure; Octavian and Antonius were doing the same in Italy.

Meanwhile Sextus Pompeius in Sicily came under attack by a fleet commanded by Q. Salvidienus Rufus and an army under Caesar's grandnephew and heir Octavian. The fleets met in the Strait. Pompeius had to prevent Salvidienus achieving a foothold on the island, for Octavian had the larger army. The battle, off the Skyllaion Cape, was once more between heavy Roman ships and the lighter Graeco-Carthaginian types. Octavian had presumably been building up a new fleet in the face of Pompeius' control of Sicily and of the supply lines from Africa and Sardinia, but the latter had the better seamen. He also may have had a larger number of ships, though the only sign of this is that Appian describes Salvidienus as commanding 'a fleet', whereas he says Pompeius had 'a large fleet'. The result was a qualified victory for Pompeius, in that he prevented Salvidienus from getting to the island and forced him to take refuge in the harbour of Balakros, near Rhegion, but this was still uncomfortably close to Sicily.[39] He was actually saved by a naval crisis which had developed in the Adriatic.

Octavian's ally, M. Antonius (Mark Antony), was at Brundisium aiming to cross the sea to Greece to tackle Brutus and Cassius, who at the time were attacking the Lykians and the Rhodians. L. Staius Murcus had adopted the method used by the Pompeian commanders during the previous trans-Adriatic confrontation, of seizing control of an island off the harbour of Brundisium and so blockading the port. He now had eighty ships and so he blocked any move by Antonius. Octavian responded by sailing all round Sicily, ignoring Pompeius, and heading straight for the Balkans. (This may well have taken Pompeius by surprise, for he did not react.) Murcus withdrew his fleet to avoid being trapped between the two forces and Antonius then got across, landing at Dyrrhachion. L. Domitius Ahenobarbus was sent to reinforce Murcus with another 50 ships and this joint force of 130 ships effectively cut communications between Italy and Greece. A third fleet, under L. Tillius Cimber, seventy strong, came forward from the Hellespont to pace the army of Brutus and Cassius by sea.[40] All forces eventually foregathered at Philippi, where the Liberators were beaten and killed.

The fleets could have made all the difference if Brutus had been able to resist the demands of his soldiers to fight. On the same day as the battle, the large fleet in the Adriatic intercepted a supply convoy when the wind failed. At their leisure the Liberators' warships forced the triremes which formed the escort to surrender, then set about sinking the transports, which were carrying two legions of infantry, which were destroyed.[41]

Both of the leaders of the Liberators were now dead, and many of the soldiers too, but their fleets was still afloat. These gathered together: one fleet under C. Cassius Parmensius included his own ships and thirty taken from Rhodes (others were burned so that the island was now virtually disarmed); thirteen ships under Clodius; a 'numerous' fleet commanded by D. Turulius, who also had a large treasury. To this force came those of their party who had survived Philippi. We are not told what happened to the fleet of Tillius (unless it was that under Turulius, with his name misspelled). Tillius' fleet certainly survived; and presumably the two fleets joined together. They then sailed round to join the main fleet of Murcus and Ahenobarbus in the Adriatic.[42]

This force should have been getting on for 300 ships strong. It split into two fleets: Murcus went to join Sextus Pompeius in Sicily, Ahenobarbus remained in the Adriatic. Murcus is said to have had eighty ships, and Ahenobarbus seventy, a total of only a few more ships than they had commanded earlier.[43] It must be concluded that many of the ships from the Aegean waters had left these fleets, probably going home to their eastern cities. Antonius, heading east, and Octavian, heading for Italy, seem to have had no difficulty in crossing the Hellespont and the Adriatic respectively. Antonius was probably now able to collect a fleet of his own in the east, from the same ships, again, which had been in the Liberators' fleets. He soon made an alliance with Kleopatra, whose Ptolemaic fleet was now once more the single greatest naval power east of the Adriatic.

The problem for the fleets of Murcus and Ahenobarbus was that they had no bases, and so they were perpetually short of supplies. They could, and did, raid the coasts of Italy, as did Sextus Pompeius, but his problem was less for he had Sicily as his base. In the east Antonius was not only allied with Kleopatra, but he was also building ships in all the dockyards he controlled. A little over a year later he came west with a fleet of 200 ships.[44] By that time Murcus had publicly joined Pompeius, who by that time had about 300 ships. Antonius was concerned at the situation in Italy, where Octavian had faced a major crisis involving Antonius' brother Lucius. When he defeated Lucius' forces the refugees fled in all directions, some to Pompeius, some to Ahenobarbus, some to Antonius. Antonius brought his fleet towards Brundisium and met Ahenobarbus' fleet on the way; by pre-arrangement Ahenobarbus now joined him. Not long afterwards Murcus succumbed to the jealousy of Pompeius' other commanders, and to Pompeius' own suspicions, and was murdered. The three survivors, Octavian, Antonius, and Sextus Pompeius, met at Misenum in the Bay of Naples and agreed a treaty, but none of them expected it to last.

Octavian had no ships to speak of before Antonius' arrival, but he was now able to develop a new naval force. The removal of Ahenobarbus' fleet into Antonius' control allowed building to take place in the Italian Adriatic ports, of which Ravenna seems to have been the most important, and the truce with Pompeius allowed him to build in the Tyrrhenian ports as well. Pompeius responded with renewed building himself.[45] He was in the weakest position, of course, despite his ships, since his land and population and resource base was the smallest of the three. He was also menaced from Africa, where M. Aemilius Lepidus had a large force of legions, and ranked, in theory, as an equal with Octavian and Antonius as one of the Triumvirate.

The truce lasted only until Octavian felt he could defeat Pompeius in battle. With two fleets and an army he tried a gigantic three-pronged attack, the western fleet sailing from Etruria under C. Calvisius Sabinus and Metrodoros, one of Pompeius' former admirals, the eastern fleet poised at Tarentum, and the army marching to Rhegion. The idea seems to have been to decoy Pompeius' main fleet to the north, bring the eastern fleet to Rhegion, and use the ships to ferry the army across the Strait to Sicily. Once he got his army ashore, Octavian could probably win. But Pompeius held the central position and his fleet and sailors were far more experienced and competent than Octavian's. Keeping a substantial fleet of forty ships at Messana, Pompeius sent his main force to stop the western fleet under Calvisius. The two fleets met near Cumae and fought a confused battle which produced casualties and damage on both sides, but no clear victory for either. Pompeius' fleet returned to Messana.[46] Octavian had gathered his legions and his fleet at Rhegion, but sensibly decided he needed clear sea superiority before launching his men on the sea.[47]

Hearing of Calvisius' fight, and that he was still coming south, Octavian went north to meet him. Pompeius, his two fleets now united, followed and attacked Octavian's fleet before it could join with Calvisius'. The superior seamanship and numbers of Pompeius' forces pushed the Roman fleet toward the shore, though some came out to fight and made a good fist of it. Pompeius retired when Calvisius' fleet came in sight.[48] Pompeius had fought two battles, in one of which he could claim a clear victory, but the enemy was still alert and awake and approaching. He was clearly in serious trouble.

He was rescued by a great storm. Pompeius' captains realized one was approaching, but most of Octavian's captains did not, though Metrodoros got his squadron out to sea to ride out the storm away from the lee shore. More than half of Octavian's fleet was destroyed, and this on top of the losses and damages suffered in the two battles. After the storm there could be no question of an invasion of Sicily.

Octavian tried again, as he had to, in 36. This time his capable friend M. Vipsanius Agrippa would organize and command, and Antonius loaned him 120 ships (out of a fleet which now counted 300 ships).[49] The strategy was similar to the first attack, but with the addition of an invasion from Africa by Lepidus. Octavian built a new fleet, but was hindered by Pompeius' repeated raids on the dockyards and ports of Italy. Agrippa, put in charge of the preparations, cut a canal to connect the Lucrine Lake to the sea, and so produced a large and safe harbour. The ships were all cataphract, but were also armoured against missiles; the oarsman received serious training. And the ships were, in the Roman tradition, built as heavy quinqueremes.[50]

So Pompeius was attacked by three fleets: Lepidus came with seventy ships from Africa and besieged Lilybaion; T. Statilius Taurus came from Tarentum with Antonius' ships, of which 102 were available (the crews of the others had died); Octavian with the western fleet moved south from Puteoli. The weather interfered again, many of Lepidus' ships being sunk, and 34 of Octavius' heavy ships (including a six) and some of the liburnians were wrecked. Taurus took heed of the weather and returned to harbour in time. Octavian at first thought he would need to delay another year, but then rebuilt his fleet and decided to press on. Everyone moved cautiously. Some of Lepidus' reinforcements were intercepted and either sunk or driven back to Africa.

Pompeius concentrated his forces, land and sea, in the northeast corner of Sicily, ignoring Lepidus, who was stuck at Lilybaion. Taurus came forward to Skylakion; Octavian to Vibo and then to the Lipari Islands, and put the fleet there under Agrippa's command. Pompeius reinforced the fleet which faced Agrippa, but the heavy Roman ships wreaked havoc among the lighter Sicilian vessels; Pompeius lost thirty ships to five of Agrippa's, and only half of Agrippa's fleet had been engaged. This was the Battle of Mylai, fought by Agrippa to draw attention away from the other forces, the eastern fleet in particular, which had now come forward to the very toe of Italy. Octavian had taken the opportunity and had moved three legions across to Tauromenium in Sicily. Pompeius, realizing the trap, sent both his land forces and his main fleet south. The fleet won a victory over Taurus' fleet, but the force on land survived. Meanwhile Agrippa landed another army at Tyndaris in the north. Lepidus brought his own forces forward from Lilybaion; the three allied forces penned Pompeius into the northeastern tip of the island.

Pompeius attempted to break out by a victory at sea, but Agrippa's ships were too powerful. He had developed a new version of the old *corvus*, called the *arpax*, 'snatcher', which seized enemy ships so that they could be boarded by the more numerous Roman marines. Each side is said to have had 300 ships in the fight off Naulochos, though this is perhaps an exaggeration. What is quite certain is the decisiveness of the victory gained by Octavian's fleet. Only three of his ships were sunk to twenty-eight of Pompeius'; but the real results came when Pompeius' survivors broke and fled. Seventeen ships got away, but the rest were trapped and either captured or wrecked or burned.[51] Pompeius escaped to Antonius, but his armies surrendered to Lepidus. Lepidus attempted to seize Sicily from Octavian, but his army would not permit it. Octavian recruited them into his own service, and Lepidus ended with nothing but the title of *pontifex maximus*.

The result of the Sicilian war was thus not merely the elimination of Pompeius and his fleet, but the elimination of one of the triumvirs as well. The Roman world was now divided between just two men; Octavian was the victor in Sicily, while Antonius was soon after defeated in his great invasion of Parthia. Another confrontation was clearly in the offing, and both men commanded great fleets.

Chapter 16

Actium

The moment of the final break between Octavian and Antonius is not easy to detect, but the failure to renew the Triumvirate – now a duumvirate – at the end of 33 is as good a point of division as any. By that time Antonius had gathered his fleet at Ephesos and was moving his legions westwards.[1] He had been building more ships and the fleet he had collected was the largest seen in the Mediterranean since that of the Great King Xerxes in 480 BC. Octavian on the other hand seems not to have been building many new ships. He had major fleets in action against Sextus Pompeius, of course, and had captured some of Pompeius' ships in the final battles. When he began the final war he is described as having 230 warships.[2] It would seem therefore that he had sorted out the best of the ships he had in 36, discarded some, and only built new vessels to bring his fleet to a reasonable total. They were accompanied by perhaps 140 liburnians. The main fleet consisted of quadriremes, quinqueremes and some sixes.

Antonius seems to have drawn a particular lesson from the sea fighting between Pompeius and Octavian. Some of his own ships had been involved, and they had been comprehensively beaten by Pompeius' fleet in the fight off Tauromenium. On the other hand, Agrippa's fleet of heavy ships at Mylai and Naulochus had won even more decisively by a combination of weight of numbers and the use of heavy and heavily-manned ships. The lesson seemed to be that Antonius needed big ships to combat Octavian's quinqueremes. In the fleet gathered at Ephesos, therefore, there were many ships of the biggest sizes, sixes to nines, in addition to the usual quinqueremes and liburnians. Of the 800 ships Antonius collected, 500 were warships, of which 200 had been provided by Kleopatra.

Antonius, therefore, had a certain advantage in numbers of ships, and, if his conclusions drawn from the Sicilian War were correct, a much greater advantage in sizes of ships. He brought the fleet around Greece to Corcyra during 32; his army, well over 100,000 men, marched across from Macedonia to camp on the shores of the Ambrakian Gulf. Octavian sent scouts to watch all this preparation, but made no move to dispute Antonius' approach. Antonius thereby appeared to be the aggressor. He was accompanied by Kleopatra, who had been demonized by Octavian's propaganda, and this rubbed off on Antonius as well, as intended. She was also a divisive presence in the camp, but since her whole fleet of 200 ships was being used, she had a perfect right to be there.

Once Antonius' fleet and army were in place, Octavian (with Agrippa, whose counsel was irreplaceable) began a strategy which resembled a siege more than an open campaign. Agrippa sailed to the small city of Methone at the south-western corner of the Peloponnese and captured it in a fight with a strong Antonian garrison.[3] This allowed Agrippa to cut that supply line, and to intercept and use any supplies which tried to pass. Antonius was compelled as a result to rely on the supplies he could gather locally and in Greece, which had to be transported laboriously over the mountains by land, a slow process which used much manpower, and guaranteed much local resentment; Sparta was soon in revolt.

Octavian's army was shifted to Epeiros without interference from the Antonian fleet. The two land forces approached each other and camped on opposite sides of the strait leading to the Ambrakian Gulf. Agrippa came north and seized Corcyra, which involved defeating a squadron of Antonius' ships, then the island of Leukas as well, which involved defeating an Antonian squadron commanded by Q. Nasidius.[4] All this further constricted Antonius' seaborne supplies, and Octavian sent some of his army across into Greece to interrupt Antonius' supply lines from the land side. Thus Antonius' fleet, now beached on the southern shores of the Gulf, and his army, were subjected to a gradually tightening ring of enemy forces, their supplies dwindled, and disease from the unhealthy situation affected them. The sheer size of Antonius' forces made them vulnerable. What was more, there was a constant leakage of men crossing the line, mainly from Antonius to Octavian; as a result the condition and even the plans of Antonius' forces were known in Octavian's headquarters. When Antonius decided to break out, Octavian knew about it in advance.

Antonius' ships had been somewhat undermanned even when he arrived, and hunger, disease, and desertion had depleted the oarsmen and sailors still further. Similarly his soldiers were wasting away. By the summer of 31 the situation had deteriorated so far that Antonius was looking not to attack Octavian or invade Italy but to escape from the trap into which his campaign had turned.

In early August, after Antonius had been on this campaign for a year, his fleet, or part of it, tried to break out. Commanded by Q. Sosius, it took on the blockading squadron while Antonius led a land force to a better camping site. Sosius' approach was concealed by mist and his ships defeated the blockaders, but in the moment of victory Agrippa arrived with the main fleet and drove Sosius back into the Gulf. Antonius, covered by this action, could clearly have got the army away, but without his fleet he would have been trapped even more decisively than in the Gulf. He returned.[5]

His options were thus steadily narrowed. Either the fleet must make another attempt to break out and so force a full-scale battle, which, if lost, would mean that his army would have to surrender or scatter, or he could march off with the army back into Macedonia and Greece, abandoning the fleet. Only the first of these offered a chance of victory, but he needed the fleet as much as he needed the army. Both were essential to his success, and indeed to his survival, and to lose one would mean he had lost the war. The decision was taken that the whole available fleet must break out.

This brought on the Battle of Actium, named for the small town on the coast near where the fighting took place. (It is also where two other great galley fights took place 1,600 years later, Preveza and Lepanto.) A timely desertion by Q. Dellius brought news of the intended break-out to Octavian, who at first thought it would be best to allow the ships to get out and then chase them. Agrippa pointed out that if a decent wind blew the Antonian forces could well get clean away, which would simply transfer the necessary battle to somewhere else. So he agreed to Agrippa's alternative, which was to force the Antonians to fight to get out.[6]

A four-day storm delayed the Antonian fleet. Antonius selected the best ships and manned them fully with oarsman, sailors, and soldiers. The rest were burned; there was no point in leaving them for Octavian. The army was put under the command of the loyal and capable P. Canidius Crassus, whose instructions must have been to save the army in any way he could.[7]

On 2 September 31, Antonius brought his fleet out of the Ambrakian Gulf. His object was to escape without fighting if he could. If it was necessary to fight, he aimed to get as many ships away as possible. He needed the afternoon breeze from the northwest to allow the ships to sail past Leukas, so the fleet came out and stopped, waiting for the breeze. Agrippa, whom Octavian had put in command, held his fleet at rest a mile away. He wanted Antonius to come out into the open sea, and he knew Antonius' own plan, so he could wait.

The fighting began in the early afternoon. The wings of Antonius' force were the stronger, for there the biggest ships were concentrated, and Agrippa necessarily had to copy that arrangement. On his left Agrippa attempted to outflank the approaching Antonian right. Agrippa had the advantage of numbers to allow him to do this, but Antonius nevertheless succeeded. Agrippa's centre had to be weakened by the concentration on the wings, while Antonius' manoeuvres left his own centre open. When the two wings were fully engaged, Kleopatra's ships, which had been held back, hoisted sails and broke through the centre. Antonius himself transferred to a quinquereme and followed.

The fighting went on until late afternoon, and even later it was not clear what the exact result was. Kleopatra and Antonius had got away with about sixty ships. Of the rest all were lost, but how many were captured, how many sunk, how many wrecked, is not known.[8] On the shore the army, which was presumably ready to march while the sea battle went on, set off by land. Octavius' army followed, and once the result of the sea battle became known, the two armies faced each other for a week while terms of surrender were agreed. The Antonian troops demanded, and got, equal treatment with Octavian's army in terms of rewards, lands, or continued service; they successfully insisted on six of Antonius' legions being taken into Octavian's service whole.

Antonius and Kleopatra escaped, but it soon became clear that their defeat was complete. The surrender of the army was the final act, though the destruction of the fleet was what had forced that surrender. They returned to Egypt. Kleopatra had an idea of fighting on with a fleet in the Red Sea, which the Ptolemies had been exploiting recently for its trading connection with India, but her ships there were burned by the Nabataeans.[9] Antonius found that his legions in Cyrenaica and Syria went over to Octavian on receiving news of the result of the battle. Egypt was, as the earlier Ptolemies had shown, a natural fortress, but it could hardly stand against a man who commanded the military and naval resources of the whole Mediterranean world.

It took Octavian nearly a year to finish the war off by invading Egypt. There was no hurry. One army marched east from Cyrenaica; a second came towards Egypt by the traditional route through Palestine and along the Sinai coastal road, accompanied by a fleet sailing along the coast. The end came on 1 August 30, eleven months after Actium. Kleopatra's fleet sailed out from Alexandria, while Antonius commanded the army to defend the city by land. But the cavalry deserted him, the infantry were beaten, and the fleet surrendered without a fight as soon as the two fleets met.[10]

Conclusion: Sea Power

Actium was the last naval battle for three centuries, though the imperial Roman regime maintained a modest navy for all that period. For the three previous centuries, which we call the Hellenistic age, there had been several competing navies in the Mediterranean. At the beginning, when Alexander gained control of the Persian fleet by conquering its home bases, and the Athenian and other Greek fleets existed, the eventual conqueror, Rome, had no ships at all. Such are the surprises of history.

Two approaches to sea power can be detected between Alexander's time and that of Octavian. One is to keep a standing fleet, with dockyards always busy building new ships and carrying out maintenance, and needing a constant supply of sailors, oarsmen, and supplies. This is expensive, needing purpose-built dockyards, ships, pay for sailors and oarsmen, supplies always available, and a major governmental bureaucratic machine capable of exacting a regular and substantial flow of taxes from a more or less submissive population, or at least a population convinced of the need for the fleet. Ptolemaic Egypt is the most obvious case, but earlier Athens had done this, and in the Hellenistic period so did Rhodes.

The second approach saw sea power as necessary only at times of war. In this case the emphasis was on making-do in emergencies, a denigration of skills, and the use of brute force. The need was for the ability to build an emergency fleet quickly, or to collect a fleet from several minor sources. This in turn permitted a lower level of taxation and a less obtrusive government, but relied on strong public support in the emergencies. There was obviously a lack of command skills in such circumstances, and reliance on heavier vessels rather than seamanship to gain the victory. This tended to result in longer periods of warfare and so heavier casualties. Rome was the exemplar here, but the rise and fall of the navies of Carthage and Syracuse suggest that this was the policy followed by those cities as well.

It will be noticed that the poor, the taxpaying peasantry and the conscripted oarsmen and soldiers, were the true casualties whichever method was adopted.

Of course, these approaches are not wholly exclusive. One of the most interesting and long-lived of the 'standing-navy' approaches was that of the Ptolemies, whose navy dominated the eastern waters of the Mediterranean for a century, but which was also enamoured of big ships, some of which were so big as to be useless in warfare. These big ships were, of course, partly built to be able to face enemy super-galleys, but they continued to be built even after the competition had been faced down or destroyed, and so they were now built as a matter of prestige. Rome, with its brute force methods, derided those big ships, at least until the very end. Only M. Antonius, in his final battle, used any ship larger than a six.

The powers which exercised sea power by maintaining a standing navy did so in order to promote their imperial power, and to promote their commercial interests. The two, as with fifth-century Athens, may well not be easy to separate. But it seems clear that the Ptolemies' main purpose was empire, while that of Rhodes, which also maintained a standing fleet, was protection of its commerce, and in both cases one of the more important effects was the suppression of piracy. Yet Rhodes developed its own mini-empire, and the Ptolemies were as commercially minded as they were imperialistic.

The Roman method of sea power was always different. A standing navy was created only in times of conflict. Instead a new fleet was called into existence whenever it was needed. At first this was an allied navy collected from seaport-cities along the Italian coasts, then, when this proved insufficient, a Roman navy was made up of ships built at Roman command and manned by conscripts, as in the two great wars with Carthage. The result of these and the succeeding wars in Greece and the Aegean was a permanent fleet but one which was laid up in times of peace, but which then rotted away through neglect. The end result was the effective destruction of all competing navies, except that of the Ptolemies, who were far off and obsequiously friendly. So, after the destruction of Macedon in 168, Rome gave up its fleet. A few ships may have been kept and maintained for future use, but when they reached the end of their useful lives, they were not replaced. This was not necessarily a decision made at the strategic level, but more a matter of neglect and an unwillingness to pay taxes.

So, instead of maintaining a standing navy of its own, Rome thrust the burden back onto its subject states. And on the whole this worked. In the Mithradatic wars it proved to be quite possible to call up those subjects' ships into a considerable fleet, and Mithradates' large fleets were defeated by these emergency collections of Greek ships, under Roman command. Much the same happened in the several attempts at suppressing piracy. It was only when the civil wars pitted Roman commanders against each other that they began to build their own ships again. Even then for a time in Sulla's wars there does not seem to have been much Roman building of ships. The navy of the civil wars was created during and by those wars.

This was empire on the cheap, but it was also a characteristically Roman approach. In land warfare the majority of soldiers in Rome's armies were always contributed by Rome's allies; in finance Roman internal taxation was abolished when the riches of the Mediterranean area flowed in during the first quarter of the second century BC. And so putting the burden on the allies to provide ships was a logical decision. It was, however, also sensible. The ships of the allies were also useful in suppressing piracy in their own waters. Rhodes is famous for this, but Athens, the Lykian states, and the Ionian cities were as active in controlling their local waters. This worked well enough until late in the second century when Kilikian piracy developed. And even that was not a really major issue until the Mithradatic wars preoccupied those local navies, leaving the seas open for pirates to roam virtually unhindered.

At the same time it was always possible to produce a new navy quickly. Frequently during this period of three centuries a major power was able to conjure up a great fleet in a very short time. Antigonos I was the first to do so, when he had the need, the money, the resources, and the skilled workmen all in one place, which in this case was Phoenicia. Later his son did the same, as did Antiochos III. Rome managed the feat more than once, as did Mithradates. Even Julius Caesar did so on a small scale when he had a squadron built at Arelate in a month. These episodes imply that the resources in wood and skills were ubiquitous all round the Mediterranean. But all these were emergency fleets, and it was such fleets which were likely to be abandoned once the war ended. Prolonged periods of peace were as lethal to navies as a major defeat in battle. The most competently managed navy of the period, that of the Ptolemies, lasted a century, but rotted away once imperial peace came and there were no competitors.

During the Hellenistic period, power at sea was either imperial or very local. Most port-cities had a few triremes to protect their immediate area and their maritime approaches. Imperial power could be projected by sea, as by the Ptolemies, but this was only done when there was a competitor to be faced. When the navies of Carthage, Macedon, and the Seleukid kingdom had been destroyed there was no need for an imperial navy at Rome – until Romans began fighting each other. But during that non-imperial period, the local squadrons continued. Then, when Octavian secured supreme power to himself alone, after 30 BC, it was no longer necessary for Rome to possess a main fleet. Most of the ships were dismantled, and only minor squadrons were maintained at several posts throughout the Mediterranean for the next three centuries – an imperial version of the preceding local system. But once again prestige was the driving force as much as anything. The flagship of the Misenum fleet, which was visited by emperors every now and again, was a six. Without an enemy to fight, this was as useless and expensive as Ptolemy Philopator's forty.

Notes

Chapter One
1. Diod. 16.89.5; Tod, *GHI* 177.
2. Arr. 1.18.5 (400); Diod. 17.29.2 (300).
3. Arr. 1.19.3–11.
4. Arr. 1.20.1; Diod. 17.22.5.
5. Diod. 17.29.1–3.
6. Curtius 3.1.19–21.
7. Arr. 1.18.6–7.
8. Arr. 2.2.1; Curtius 3.3.1.
9. Arr. 2.2.2–3; Curtius 4.1.36.
10. Arr. 2.2.4.
11. Curtius 4.1.37 and 5.3; Arr. 2.5.7.
12. Arr. 2.13.4.
13. Arr. 2.13.7.
14. Curtius 4.5.14–22.
15. Arr. 2.12.2; Curtius 4.1.27–30; Diod. 17.48.1–5.
16. Arr. 2.13.7 and 15.6.
17. Arr. 2.196–20.2.
18. Arr. 2.22.2–5; Diod. 17.423–4.
19. Arr. 3.7.3–7.
20. Arr.3.5.5; Curtius 4.8.4.
21. Arr. 3.6.3.
22. Arr. 2.17.1–4.
23. Arr. 6.11.
24. Arr. 7.19.3; Curtius 10.16.
25. Justin 13.5.7.

Chapter Two
1. *Inscriptiones Graecae* II(2) 1627, 266–8, and 1629, 805–12.
2. Diod. 18.10.2.
3. Morrison, *GROW* 15–16.
4. J.S. Morrison and R.T. Williams, *Greek Oared Ships*, Cambridge 1968, ch.8 (by D.J. Blackman).
5. J. S. Morrison, 'Hyperesia in naval contexts', *JHS* 104, 1984, 48–59.
6. Casson, *Ships* 90 and 120.
7. Ps.-Demosthenes 17.19.
8. Tod, *GHI* 200.
9. Strabo 10.1.8.
10. Diod. 18.8.1.
11. Diod. 18.15.8–9.
12. Morrison, *GROW* 18.

13. *Inscriptiones Graecae* II(2), 38 and 493; N.G. Ashton, 'The *Naumachia* near Amorgos in 322 BC', *ABSA* 72, 1977, 1–11.
14. Marmor Parium, *FGrH* 239 B, para. 8.
15. Peter Garnsey, *Famine and Food Supply in the Graeco-Roman World*, Cambridge 1988, 156–157.
16. Ashton (note 13).
17. Plutarch, *Moralia* 338a, and *Demetrios* 11.3.

Chapter Three
1. Diod. 18.33.1–6; Arr., *Successors*, *FGrH* 156, 9.28–29.
2. Arr., *Successors*, *FGrH* 156, 9.39.
3. Diod. 18.69.1.
4. Diod. 18.72.2–9.
5. Diod. 18.73.1.
6. Diod. 18.75.1.
7. Diod. 18.63.6; Polyainos 4.6.9.
8. W.W. Tarn, *Antigonos Gonatas*, Oxford 1913, 72.
9. Diod 19.58.1–5.
10. Diod 19.58.5–6.
11. Diod. 18.62.3–4.
12. Diod. 19.60.4.
13. Diod. 19.62.6–8.
14. Diod. 19.64.4–7.
15. Diod. 19.62.8; Morrison, *GROW* 21.
16. Diod. 19.69.3.
17. Diod. 19.73.6–10.
18. Diod. 19.75.7–8.
19. Diod. 19.77.2–5 and 7.
20. Diod. 20.19.4–5.
21. Diod. 20.27.1–2.
22. Diod. 20.37.1–2.

Chapter Four
1. Plut., *Demetrios* 8; Diod. 20.45.4.
2. Plut., *Demetrios* 8.3.
3. Diod. 20.47.1.
4. Diod. 20.49.2; Plut., *Demetrios* 16.1.
5. Diod. 19.78.4.
6. Plut., *Demetrios* 8.3–4; Polyainos 4.7.6.
7. Plut., *Demetrios* 9; Diod. 20.45.2–46.2.
8. Diod. 20.48.3; Plut., *Demetrios* 9.
9. Diod. 20.46.4; Plut., *Demetrios* 10.
10. Diod. 20.46.5.
11. John H. Pryor, *Geography, technology and war: Studies in the maritime history of the Mediterranean, 649–1571*, Cambridge 1988, ch 1.
12. Diod. 20.46.6; Plut., *Demetrios* 15.
13. Diod., 20.47.1.
14. Plut., *Demetrios* 16.1–2; Polyainos 4.17.7; H. Hauben, 'Fleet Strength at the Battle of Salamis (306 B.C.)', *Chiron* 6, 1976, 1–5.
15. Diod 20.47.1–4 and 48.1–4.
16. Diod 20.49.1–2; Plut., *Demetrios* 16.
17. Diod 20.49.5–52.5; Plut., *Demetrios* 16; Polyainos 4.7.7.
18. Diod. 20.52.6; Plut., *Demetrios* 16.
19. Diod. 20.73.1–2.
20. Diod. 20.72.2–76.7.

21. Diod. 20.81.1–82.4; H. Hauben, 'Rhodes, Alexander and the Diadochi from 333/332 B.C. to 304 B.C.', *Historia* 26, 1977, 307–339.
22. Diod. 20.82.4.
23. Diod. 20.82.5–88.9 and 91.1–98.9.
24. Diod. 20.99.1–3.
25. Diod. 20.100.5–6; Plut., *Demetrios* 23.

Chapter Five
 1. Diod. 20.107–111, *passim*.
 2. Diod. 20.112.
 3. Plut., *Demetrios* 30.1–2.
 4. Plut., *Demetrios* 30.3–31.1.
 5. Plut., *Demetrios* 31.2–32.2.
 6. Plut., *Demetrios* 33.
 7. Plut., *Demetrios* 35.
 8. Plut., *Demetrios* 43.
 9. Plut., *Demetrios* 44–46.
10. Plut., *Demetrios* 46–49.
11. J. Seibert, 'Philokles, Sohn des Apollodoros, König der Sidonier', *Historia* 19, 1970, 337–351; H. Hauben, 'Philocles, King of the Sidonians and General of the Ptolemies', *Studia Phoenicia* 5, 1987, 413–42.
12. Aelian, *Varia Historia* 6.12; Pliny, *Natural History* 7.207.
13. Morrison, *GROW* 34–35; Casson, *Ships* ch. 6; W.W. Tarn, *Hellenistic Military and Naval Developments*, Cambridge 1930, ch. 3.
14. Plut., *Demetrios* 31–32.
15. Plut., *Demetrios* 43; Casson, *Ships* 112–114; Morrison, *GROW* 233.
16. Plut., *Demetrios* 20.
17. Plut., *Demetrios* 43.
18. Athenaios 5.203c.

Chapter Six
 1. Appian, *praef.* 10.
 2. Athenaios 5.203c.
 3. J.D. Grainger, *The Syrian Wars*, Leiden 2010.
 4. G. Holbl, *A History of the Ptolemaic Empire*, London 2001, 23–24.
 5. Note 11, ch. 5.
 6. See the details in R. Bagnall, *The Administration of the Ptolemaic Possessions outside Egypt*, Leiden 1976.
 7. H. Hauben, *Callicrates of Samos. A Contribution to the Study of the Ptolemaic Admiralty*, *Studia Hellenistica* 16, Leiden 1970.
 8. Pausanias 1.1.1; Strabo 9.1.21.
 9. Bagnall (note 6), 81 and 135–6.
10. J.D. Grainger, *The Cities of Pamphylia*, Oxford 2009.
11. Dionysios of Halicarnassos, *Roman History* 14.1; Livy, *Per.* 14.
12. P.M. Fraser, *Ptolemaic Alexandria*, Oxford 1972, vol. I, 152 and 522.
13. N.L. Grach, 'A New Monument of the Hellenistic Age from Nymphaeum', *Vestnik Dreveistoria* 1, 1984, 81–88; L. Busch, 'The Isis of Ptolemy II Philadelphos', *The Mariner's Mirror* 71, 1985, 129–151; Morrison, *GROW*, 207–214; W.M. Murray, 'A trireme named Isis: the graffito from Nymphaeum', *International Journal of Nautical Archaeology* 30, 2000, 250–256.
14. S.M. Burstein, 'Ivory and Ptolemaic Exploration of the Red Sea: the Missing Factor', *Topoi* 6, 1996, 799–807; H.H. Scullard, *The Elephant in the Greek and Roman World*, London 1974, 126–137.
15. Athenaios 5.209e; Pausanias 1.29.1.

16. Recent discussions: N.G.L. Hammond and F.W. Walbank, *A History of Macedonia*, Oxford 1988, vol. III, app 4, 'The battles of Cos and Andros'; G. Reger, 'The Date of the Battle of Kos', *American Journal of Antiquity*, 10, 1985, 155–177.
17. Polyainos 5.18.
18. G. Reger, 'The Political History of the Kyklades, 260–200 B.C.', *Historia* 43, 1994, 32–69.
19. P. Bruneau, *Recherches sur les cultes de Délos à l'époque hellénistique et l'époque impériale*, Paris 1970.
20. M.M. Austin, *The Hellenistic World*, 2nd ed., Cambridge 2006, no. 266.
21. Note 16.
22. Reger (note 18).
23. Bagnall (note 9).

Chapter Seven

1. Morrison, *GROW* 3–4.
2. Thiel (1), 20–26.
3. R.J.A. Talbert, *Timoleon and the Revival of Greek Sicily, 344–317 B.C.*, Cambridge 1974.
4. Plut., *Agis* 3.2; Diod. 16.62.4 and 63.2; Pliny, *Natural History* 3.98.
5. Plut., *Timoleon* 11.5, 13.4, and 17.2–3; Dionysios of Halikarnassos 69.3.
6. Diod. 16.68.4–8.
7. Plut., *Timoleon* 25.1.1.
8. Plut., *Timoleon* 34; Diod. 16.82.3.
9. Diod. 19.71.
10. Diod. 19.102.8.
11. Diod. 19.103.4.
12. Diod. 19.106.2–3.
13. Diod. 19.107.2; cf. H.J.W. Tillyard, *Agathocles*, Cambridge 1908, 88.
14. Diod. 20.5.1–8.1 and 9.1–2.
15. Diod. 20.32.3–5 and 61.7.
16. Diod. 20.61.5–62.1.
17. Diod. 20.101.
18. Diod. 20.64.1 and 105.3; Livy 10.1–2; Pausanias 4.36.2; Athenaios 13.605d.
19. Diod. 21.2.13.
20. Diod. 1.4.
21. Livy 9.28.7; Diod. 19.101.3.
22. Livy 9.30.14.
23. Livy 9.38.2–3; Diod. 19.65.7.
24. Livy 9.43.26; Polybios 3.22–26.
25. Appian, *Samnite Wars* 7.1; Cassius Dio 2.22–26.
26. Diod. 21.2.2.
27. Diod. 21.4 and 8.
28. Diod. 21.16.1.
29. Diod. 22.8.5.
30. Diod. 22.8.1–5.
31. Polybios 3.25/1–9; Diod. 22.7.
32. Appian, *Samnite Wars* 7.2.
33. Plut., *Pyrrhos* 24.

Chapter Eight

1. Polybios 1.7.6–8 and 9.3–9; Diod. 22.13; Zonaras 8.6.
2. Polybios 1.10.1–2.
3. Diod. 22.13.6–7; Polybios 1.10.1.
4 . Diod. 23.2; Polybios 1.11.10–11.
5. Thiel (1), 33.
6. Polybios 1.17.4–6.
7. Polybios 1.20.7.

8. *Ineditum Vaticanum* 4.
9. Polybios 1.20.9.
10. Polybios 1.20.10–21.3.
11. Polybios 1.21.11.
12. Zonaras 8.11.
13. Thiel (1), 171–178; Lazenby 62–66.
14. Polybios 1.21.4.
15. Polybios 1.21.3–8.
16. Polybios 1.1.9–22.1.
17. Polybios 1.22.2–3.
18. Polybios 1.23.
19. Zonaras 8.11; Valerius Maximus 5.1; Florus 1.18.16.
20. Polybios 1.24.5–7; Zonaras 8.12.
21. Polybios 1.24–25.4; Zonaras 8.12; Polyainos 8.20.
22. Discussions in Lazenby 81–84; Thiel (1), 84–85; W.W. Tarn, 'The Fleets of the First Punic War', *JHS* 27, 1907, 48–60.
23. Polybios 1.26.10–28.14; G.K. Tipps, 'The Battle of Ecnomus', *Historia* 34, 1985, 432–466.
24. Polybios 1.29.1; Zonaras 8.12.
25. Polybios 1.29.1–5.
26. Polybios 1.29.9.
27. Polybios 1.36.10–11.
28. Polybios 1.37.1–7.
29. Polybios 1.38.5.
30. Polybios 1.38.6–9; Diod. 23.18.3–5.
31. Polybios 1.39.2–6.
32. Polybios 1.39.13–15.
33. Polybios 1.42.7 and 44.1–7.
34. Polybios 1.46.4–47.10.
35. Polybios 1.53.1–2.
36. Polybios 1.53.1–2.
37. Polybios 1.53.5–54.8; Diod. 24.1.7–9.
38. Zonaras 8.16.
39. Polybios 1.56.2–4, 10–11.
40. Polybios 1.59.6–8.
41. Polybios 1.59.9–12.
42. Polybios 1.60.1–3.
43. Polybios 1.60.4–61.7.
44. Polybios 1.62.1–9.
45. Polybios 1.63.7.

Chapter Nine
1. Polybios 1.79.2–4.
2. Polybios 1.88.8–12.
3. Polybios 1.73.2.
4. Polybios 2.11.1.
5. Cf. Thiel (2), 343–344.
6. Livy 21.49.2–4.
7. Polybios 3.33.4.
8. Polybios 3.41.2.
9. Livy 21.42.2.
10. Livy 21.49–50.
11. Livy 21.51.
12. Livy 21.51.6–7; Polybios 3.61.9–14.
13. Livy 21.32.1–4; Polybios 3.49.3–4.
14. Livy 22.11.7–10.

15. Polybios 3.76.1–3.
16. Polybios 3.95.1–2.
17. Polybios 3.95.5; Livy 22.19.3.
18. Polybios 3.95.5–96.6; Livy 22.19.3–20.2.
19. Polybios 3.96.1–14.
20. Livy 22.19.3–20.2.
21. Livy 22.56.10.
22. Livy 23.41.8.
23. Livy 23.34.1–9.
24. Livy 23.32.8–12; 40–41.7.
25. Livy 24.40.2–12.
26. Livy 24.11.5–6; Thiel (2), 74–76.
27. Livy 24.27.5; Polybios 8.1.7.
28. Livy 24.27.8.
29. Polybios 8.4.1–6.6; Livy 24.34.
30. Livy 24.35.3–36.4 and 7.
31. Livy 25.25.11.
32. Livy 25.25.13.
33. Livy 25.27.2–12.
34. Livy 25.30.7–12.
35. Livy 26.20.
36. Livy 26.39.
37. Livy 26.20.8.
38. Polybios 10.14–15.3; Livy 26.47.
39. Livy 27.22.5.
40. Livy 27.29.10.
41. Livy 27.4.6.
42. Livy 28.1–8.
43. Livy 28.16.15.
44. Livy 28.17.13–14.
45. Livy 28.30.
46. Livy 28.36–37 and 46.
47. Livy 28.46.
48. Livy 29.5.5–13.
49. Livy 30.2.1–6.
50. Livy 30.19.5–6; 25.11.
51. Polybios 14.10.2–12; Livy 30.10.
52. Livy 30.24.
53. Livy 30.43.11.

Chapter Ten
1. Polybios 7.9; Livy 23.33.10–11.
2. Livy 23.48.3.
3. Livy 24.40.
4. Polybios 5.109.1–3; cf. Casson, *Ships* 125–127.
5. Polybios 5.109.4–110.3.
6. Livy 23.38.9–10.
7. Livy 24.40.1–9
8. F.W. Walbank, *Philip V of Macedon*, Cambridge 1940, 12–13.
9. Polybios 5.43.1.
10. Polybios 5.62.2–3.
11. Athenaios 5.203c-e; A.W. Sleeswyk and F. Meijer, 'Launching Philopator's 'Forty'', *IJNA* 23, 1997, 115–118.
12. V. Gabrielsen, *The Naval Aristocracy of Hellenistic Rhodes*, Aarhus 1997.
13. Strabo 14.2.5.

14. Polybios 4.37.8–52.10.
15. Polybios 5.21.1–10.
16. Polybios 7.11.9.
17. Livy 28.5.1.
18. Polybios 8.13–14.
19. Livy 26.24; Polybios 9.39.3 and 11.26.3.
20. Livy 26.24.15–16; Polybios 9.39.2.
21. Livy 26.26.2; Polybios 9.39.2.
22. Livy 22.8.10; Polybios 9.42.5.
23. Livy 27.31.1–3.
24. Livy 27.30.16.
25. Livy 27.15.7.
26. Polybios 22.8.10.
27. Livy 28.5.1–6.12.
28. Livy 28.7.16–8.6.
29. Livy 28.8.7–10.
30. Livy 28.8.14.
31. Livy 28.30.1–16.
32. Livy 28.7.14.
33. Livy 29.12.1; Polybios 11.4.6.
34. Livy 29.12.3.
35. Livy 29.12.3–16.

Chapter Eleven
 1. Polybios 18.54.8–12; Diod. 28.1.
 2. Polybios 13.5.1–3; Polyainos 5.17.2.
 3. Diod. 27.3.
 4. Appian, *Macedonian Wars* 4.1; Polybios 3.2.8.
 5. Polybios 15.25–36.
 6. Polybios 15,20.1–8; Livy 31.14.5; Appian, *Macedonian Wars* 4.1; Justin 30.2.8.
 7. Polybios 15.22–23.
 8. Polybios 16.2.9 and 7.1.
 9. Polybios 16.10.1; 14.5; 15,1–8.
10. Polybios 16.2–9.
11. Polyainos 4.18.2.
12. Livy 31.15.5.
13. Livy 31.16.4–17.11; Polybios 16.19.3 and 30–33.
14. Polybios 16.25.2–3.
15. Livy 31.14.2–3.
16. Livy 31.23.1–12.
17. Livy 31.44–45.
18. Livy 31.45.6–11.
19. Polybios 16.7.1; Livy 31.46.8.
20. Livy 32.16.1–17; Zonaras 9.16.2.
21. Polybios 41a.1–2; Livy 33.19.10–11.
22. Livy 33.20.4–5; Hieronymos, *In Danielam* 16.15.
23. Livy 33.38.
24. J.D. Grainger, 'Antiochos III in Thrace', *Historia* 15, 1996, 329–345.
25. Livy 34.29.1.
26. Livy 35.43.2–3 and 6.
27. Livy 35.22.2 and 36.6–3.1.
28. Livy 35.24.7 and 39.5.
29. Livy 36.20.5–21.1.
30. Livy 36.3.4–6 and 42.1–2; Appian, *Syrian Wars* 22.

31. Livy 36.42.6–7.
32. Livy 36.43.1–3.
33. Livy 36.43.12.
34. Livy 36.45.5–8.
35. Livy 36.43.8–45.4; Appian, *Syrian Wars* 22.
36. Livy 36.43.8.
37. Livy 37.2.2–3, and 10.4.2–5.
38. Livy 37.9.6.
39. Livy 37.1–11.14.
40. Livy 37.11.15–12.6.
41. Livy 37.12.10–13.6.
42. Livy 37.13.12.
43. Livy 37.14.3.
44. Livy 37.14.1–4.
45. Livy 37.15.
46. Livy 37.22.3–5.
47. Livy.37.23.5.
48. Livy 37.23.6–24.10.
49. Livy 37.24.11–13.
50. Livy 37.26.5–30.10; Appian, *Syrian Wars* 29.
51. Livy 37.31.5–7.
52. Livy 37.31.2.
53. Livy 37.45.2.
54. Livy 38.39.1–6; Polybios 21.43.1–3.

Chapter Twelve
 1. Livy 30.43.1.
 2. Polybios 18.44.6; Livy 33.30.
 3. Polybios 21.42.13.
 4. Livy 36.42.2.
 5. Livy 42.27.1.
 6. Livy 40.18.7.
 7. Livy 41.9.3.
 8. Livy 42.27.2–3 and 7.
 9. Appian, *Punic Wars* 75.
10. Strabo 14.1.38.
11. Livy 40.18.7–8.
12. Livy 40.28.7.
13. Livy 42.46.5.
14. Livy 41.1.2–4.
15. Livy 42.27.2–3.
16. Appian, *Punic Wars* 75.
17. Livy 42.48.7–8.
18. Livy 42.48.7–8.
19. Livy 42.48.9.
20. Livy 43.9.5–6.
21. Livy 42.56.6–7; Polybios 27.7.14–16.
22. Livy 44.8.12.
23. Polybios 27.7.14.
24. Livy 43.3.5–13 and 7.5–8.10.
25. Livy 44.20.2.
26. Livy 44.30.13–14; Appian, *Illyrian Wars* 9.
27. Livy 44.28.1–29.5.
28. Livy 44.30.1.

29. Livy 44.35.8 and 13; 46.3.
30. Livy 44.42.5–7.
31. Livy 45.35.3; Plut., *Aemilius Paullus*.
32. Polybios 36.5.9.
33. Polybius 29.27.9.
34. Polybios 31.2.11.
35. Appian, *Syrian Wars* 46; Livy/Julius Obsequens 15.
36. Polybios 38.16.2–3.
37. Livy, *Per.* 48.
38. Appian, *Punic Wars* 99.
39. Appian, *Punic Wars* 123.
40. Appian, *Punic Wars* 121.
41. Appian, *Punic Wars* 122–123.

Chapter Thirteen
1. J.D. Grainger, *The League of the Aitolians*, Leiden 1999.
2. Livy 8.14.
3. Tod, *GHI* 200.
4. Diod. 21.3.
5. Strabo 5.3.5.
6. *Inscriptiones Graecae* II(2) 1629.
7. Diod. 20.82.5–83.1.
8. Polyainos 4.6.18.
9. Livy 37.11; Appian, *Syrian Wars* 24.
10. Polybios 30.31.
11. Strabo 14.
12. E.g., M.M. Austin, *The Hellenistic World*, 2nd ed., Cambridge 2003, no. 113.
13. Strabo 14.5.2.
14. *Ibid.*
15. Strabo 14.3.2.
16. Athenaios 6.273a; Diod. 33.28A; Justin 38.8.8; Strabo 14.5.2.
17. Orosius 5.13.3; Livy, *Per.* 68.
18. Livy, *Per.* 68.
19. *Corpus Inscriptionum Latinarum* I(2) 2662.
20. Tacitus, *Annals* 12.62.
21. M. Hassall, M. Crawford, and J. Reynolds, 'Rome and the Eastern Provinces at the end of the Second Century B.C.', *JRS* 64, 1974, 195–220.

Chapter Fourteen
1. Appian, *Mithridatic Wars* 13 and 17.
2. Dittenberger, *Orientis Graeci Inscriptiones Selectae* 552–554.
3. Appian, *Mithridatic Wars* 17.
4. W.S. Ferguson, *Hellenistic Athens*, London 1911, 377; C. Habicht, *Athens from Alexander to Antony*, Cambridge MA 1999, 283–285.
5. Appian, *Mithridatic Wars* 29.
6. Appian, *Mithridatic Wars* 56.
7. Appian, *Mithridatic Wars* 12 and 14.
8. Appian, *Mithridatic Wars* 17.
9. Appian, *Mithridatic Wars* 19.
10. Appian, *Mithridatic Wars* 24.
11. Appian, *Mithridatic Wars* 25.
12. Appian, *Mithridatic Wars* 28.
13. Appian, *Mithridatic Wars* 28; Pausanias 3.23.3–5.
14. Appian, *Mithridatic Wars* 29.
15. Athenaios 5.214d–215b; F. Durrbach, *Choix d'inscriptions de Délos*, Paris 1921, 235–236.

16. Appian, *Mithridatic Wars* 51.
17. Plut., *Lucullus* 2.3–5.4; Appian, *Mithridatic Wars* 56.
18. Plut., *Lucullus* 3.4–5.4; Appian, *Mithridatic Wars* 56.
19. Plut., *Lucullus* 3.4–7.
20. Appian, *Mithridatic Wars* 56.
21. Plut., *Sulla* 22.
22. Plut., *Sulla* 24.
23. Appian, *Civil Wars* 1.84; Livy, *Per.* 85; Velleius Paterculus 2.25.1–2.
24. Appian, *Mithridatic Wars* 63.
25. Appian, *Mithridatic Wars* 78.
26. Appian, *Mithridatic Wars* 63.
27. Appian, *Mithridatic Wars* 56; Plut., *Lucullus* 2.5 and 3.2.
28. Appian, *Mithridatic Wars* 63.
29. Strabo 14.5.7.
30. Cicero, *In Verrem* II.1.21 and 4.22.
31. Strabo 14.3.2 and 5.2.
32. Plutarch, *Sertorius* 8.
33. Plutarch, *Sertorius* 12 and 21.
34. Sallust, *Histories* frag. III 2M; Cicero, *In Verrem* II.2.8 and 3.214–215.
35. Sallust, *Histories* frag. 5–6M.
36. Florus 3.7.2 and 7; Appian, *Civil Wars* 6.1; Diod. 40.1.
37. Plut., *Lucullus* 13.4.
38. Plut., *Sertorius* 24.3.
39. Plut., *Lucullus* 11.7; Memnon 43.1.
40. Memnon 37M; Appian, *Mithridatic Wars* 71.
41. R.K. Sherk, *Rome and the Greek East to the Death of Augustus*, Cambridge 1988, no. 71.
42. Appian, *Mithridatic Wars* 76; Plut., *Lucullus* 11.6; Memnon 28.1.
43. Appian, *Mithridatic Wars* 77; Plut., *Lucullus* 12.2–3.
44. Appian, *Mithridatic Wars* 77; Plut., *Lucullus* 12.3–5.
45. Memnon F33.1–2.
46. Appian, *Mithridatic Wars* 76.
47. Memnon 33.1.
48. Plut., *Pompeius* 24; Sherk (note 41), no. 71C.
49. Livy, *Per.* 99; Plut., *Pompeius* 27; Cassius Dio 36.31.
50. Cicero, *On the Command of Cn. Pompeius*; Plut., *Pompeius* 25–26; Cassius Dio 35.
51. Appian, *Mithridatic Wars* 92–93.
52. Plut., *Pompeius* 24.
53. Plut., *Caesar* 2; Suetonius, *Julius* 4.74.
54. Appian, *Mithridatic Wars* 94.
55. Appian, *Mithridatic Wars* 95.
56. Appian, *Mithridatic Wars* 96; Plut., *Pompeius* 28; Strabo 14.3.3.

Chapter Fifteen

1. Caesar, *Civil War* 1.25; Appian, *Civil Wars* 2.40.
2. Caesar, *Civil War* 1.34.
3. Caesar, *Civil War* 1.36.
4. Caesar, *Civil War* 1.56.
5. Caesar, *Civil War* 2.3.
6. Caesar, *Civil War* 2.4–7.
7. Lucan, *Civil War* 3.509–762; Morrison, *GROW* 129–132.
8. Appian, *Civil Wars* 2.55.
9. Appian, *Civil Wars* 2.49.
10. Caesar, *Civil War* 2.23; Appian, *Civil Wars* 2.44.
11. *Ibid.*
12. Caesar, *Civil War* 2.18 and 21.

13. *African War* 64.
14. Caesar, *Civil War* 3.3; Plut., *Pompeius* 64.
15. Appian, *Civil Wars* 2.49.
16. Caesar, *Civil War* 3.5.
17. Caesar, *Civil War* 3.6; Appian, *Civil Wars* 2.54.
18. Caesar, *Civil War* 3.26.
19. Caesar, *Civil War* 3.39–40.
20. Caesar, *Civil War* 3.23–25.
21. Caesar, *Civil War* 3.40.
22. Caesar, *Civil War* 3.101.
23. Appian, *Civil Wars* 2.88; Suetonius, *Julius* 63.
24. Caesar, *Civil War* 3.106.
25. Caesar, *Civil War* 3.111.
26. *Alexandrian War* 1.1 and 9.
27. *Alexandrian War* 1.11.
28. *Alexandrian War* 1.12–13.
29. *Alexandrian War* 1.13–16.
30. Appian, *Civil Wars* 2.87.
31. *African War* 1.
32. *African War* 2.
33. *African War* 62.
34. *African War* 96; Appian, *Civil Wars* 2.97.
35. Appian, *Civil Wars* 3.4 and 4.84.
36. Appian, *Civil Wars* 4.60–62.
37. Appian, *Civil Wars* 4.65–73.
38. Appian, *Civil Wars* 4.74 and 82.
39. Appian, *Civil Wars* 4.85.
40. Appian, *Civil Wars* 4.86 and 102.
41. Appian, *Civil Wars* 4.115–116.
42. Appian, *Civil Wars* 5.2.
43. Appian, *Civil Wars* 5.25–26.
44. Appian, *Civil Wars* 5.55.
45. Appian, *Civil Wars* 5.77–78.
46. Appian, *Civil Wars* 5.81–83.
47. Appian, *Civil Wars* 5.84.
48. Appian, *Civil Wars* 5.85–86.
49. Appian, *Civil Wars* 5.93.
50. Cassius Dio 48.2–4, 50–52.
51. Appian, *Civil Wars* 5.98–121.

Chapter Sixteen
1. Plut., *Antony* 56.
2. Orosios 6.19; Florus 2.21; Morrison, *GROW* 162–165.
3. Cassius Dio 50.11.3.
4. Cassius Dio 50.13.5.
5. Cassius Dio 50.14.1–2.
6. Cassius Dio 50.14.4–15.3.
7. Cassius Dio 50.15, 4–23.1–3, 31.1–2.
8. Plut., *Antony* 65–68; Cassius Dio 50.29, 31–33; Morrison, *GROW* 165–170.
9. Plut., *Antony* 69.3.
10. Plut., *Antony* 70; Cassius Dio 51.9–10.

Abbreviations and Bibliography

Abbreviations
The following abbreviations have been used in the Notes and the Bibliography:

ABSA – Annual of the British School at Athens
AJAH – American Journal of Ancient History
Arr. – Arrian
Casson, *Ships* – L. Casson, *Ships and Seamanship in the Ancient World* (Princeton NJ, 1986)
Diod. – Diodoros of Halicarnassos, *Histories*
FGrH – F. Jacoby, *Die Fragmente der griechischen Historiker* (Leiden, 1957–1958)
IJNA – International Journal of Nautical Archaeology
JHS – Journal of Hellenic Studies
JRS – Journal of Roman Studies
Lazenby – J.F. Lazenby, *The First Punic War* (London, 1996)
MM – The Mariner's Mirror
Morrison, *GROW* – J.S. Morrison, *Greek and Roman Oared Warships* (Oxford, 1996)
Plut. – Plutarch
Thiel (1) – J. Thiel, *A History of Roman Sea Power before the Second Punic War* (Amsterdam, 1954)
Thiel (2) – J. Thiel, *Studies on the History of Roman Sea Power in Republican Times* (Amsterdam, 1946)
Tod, *GHI* – M.N. Tod, *Greek Historical Inscriptions*, vol. II (Oxford, 1948)

Bibliography
The notes have been restricted to identifying ancient sources, and to the essential modern books and articles. The following are modern materials which have been useful as well (see also the abbreviations):

N.G. Ashton, 'The *Naumachia* near Amorgos in 322 B.C.', *ABSA* 72 (1977), 1–11
M.M. Austin, 'Hellenistic Kings, War, and the Economy', *Classical Quarterly* 36 (1986), 450–461
M.M. Austin, *The Hellenistic World*, second ed. (Cambridge, 2003)
R.S. Bagnall, *The Administration of the Ptolemaic Possessions outside Egypt* (Leiden, 1976)
L. Basch, 'The *Isis* of Ptolemy II Philadelphus', *MM* 71 (1985), 129–151
L. Basch, *Le musée imaginaire de la marine antique* (Athens, 1987)
G.F. Bass, 'Greek and Roman harbourworks', *A History of Seafaring based on Underwater Archaeology* (London, 1982), 88–101
R.M. Berthold, 'Lade, Pergamum and Chios, Operations of Philip V in the Aegean', *Historia* 24 (1975), 150–163
R.M. Berthold, *Rhodes in the Hellenistic Age* (London, 1984)
R.A. Billows, *Antigonos the One-Eyed and the Creation of the Hellenistic State* (California, 1990)
D. Blackman, 'Progress in the Study of ancient shipsheds: a Review', in C. Beltrane (ed.) *Boats, Ships and Shipyards, Proceedings of the Ninth International Symposium on Boat and Ship Archaeology, Vienna 2000* (Oxford, 2003), 81–90

P. Bruneau, *Recherches sur les cultes de Délos à l'époque hellénistique et l'époque impériale* (Paris, 1970)

S.M. Burstein, 'Ivory and Ptolemaic Exploration of the Red Sea, the Missing Factor', *Topoi* 6 (1996), 799–807

J.M. Carter, *The Battle of Actium, the Rise and Triumph of Augustus Caesar* (London, 1970)

L. Casson, *The Ancient Mariners* (London, 1960)

L. Casson, *Ships and Seafaring in Ancient Times* (London, 1994)

L. Casson and J.R. Steffy, *The Athlit Ram* (Texas, 1991)

F. Durrbach, *Choix d'Inscriptions de Delos* (Paris, 1921)

R.M. Errington, 'Diodorus Siculus and the Chronology of the Early Diadochoi 320–311 B.C.', *Hermes* 105 (1977), 478–505

W.F. Ferguson, *Hellenistic Athens* (London, 1911)

P.M. Fraser, *Ptolemaic Alexandria* (Oxford, 1972)

H. Frost, *Under the Mediterranean* (London, 1963)

H. Frost, 'The Punic Wreck in Sicily', *IJNA* 3 (1974), 35–54

V. Gabrielsen, *The Naval Aristocracy of Hellenistic Rhodes* (Aarhus, 1997)

R. Gardiner (ed.), *The Age of the Galley* (London, 1995)

Y. Garlan, 'Signification historique de la Piraterie Grecque', *Dialogues d'Histoire Ancienne* 4 (1978), 1–16

R. Garland, *The Piraeus from the Fifth to the First Century B.C.* (Cornell, 1987)

P. Garnsey, *Famine and Food Supply in the Graeco-Roman World* (Cambridge, 1988)

A. Goldsworthy, *The Punic Wars* (London, 2000)

N.L. Grach, 'A New Monument of the Hellenistic Age from Nymphaeum', *Vestnik Dreveistoria* 1 (1984), 81–88

J.D. Grainger, *Hellenistic Phoenicia* (Oxford, 1991)

J.D. Grainger, 'Antiochos III in Thrace', *Historia* 45 (1996), 329–345

J.D. Grainger, *The League of the Aitolians* (Leiden, 1999)

J.D. Grainger, *The Roman War of Antiochos the Great* (Leiden, 2002)

J.D. Grainger, *The Cities of Pamphylia* (Oxford, 2009)

J.D. Grainger, *The Syrian Wars* (Leiden, 2010)

E.S. Gruen, *The Hellenistic World and the Coming of Rome* (California, 1984)

C. Habicht, *Athens from Alexander to Antony* (Cambridge MA, 1997)

N.G.L. Hammond and G.L. Griffith, *A History of Macedonia* vol. III (Oxford, 1988)

M. Hassall, M. Crawford, and J. Reynolds, 'Rome and the Eastern Provinces at the End of the Second Century B.C.', *JRS* 64 (1974), 195–220

H. Hauben, *Callicrates of Samos: A Contribution to the Study of the Ptolemaic Admiralty*, *Studia Hellenistica* 18 (Leuven, 1970)

H. Hauben, 'The Expansion of Macedonian Sea Power under Alexander the Great', *Ancient Society* 7 (1976), 79–105

H. Hauben, 'Fleet Strength at the Battle of Salamis (306 B.C.)', *Chiron* 6 (1976), 1–5

H. Hauben, 'Rhodes, Alexander and the Diadochi from 333/332 to 304 B.C.', *Historia* 26 (1977), 307–339

H. Hauben, 'Philocles, King of the Sidonians and General of the Ptolemies', *Studia Phoenicia* 5 (1987), 413–427

G. Holbl, *A History of the Ptolemaic Empire* (London, 2001)

M. Holleaux, 'Remarques sur le Papyrus de Gourob, *Bulletin de Correspondence Hellenique* 30 (1906), 330–348

H. Hurst and L.E. Stager, 'A metropolitan landscape: the Late Punic port of Carthage', *World Archaeology* 9 (1978), 334–346

E.G. Huzar, *Mark Antony, a Biography* (Minneapolis, 1978)

C. Jacob and F. de Polignac, *Alexandria, third century B.C., the Knowledge of the World in a single city* (Alexandria, 2000)

R. Kallet-Marx, *Hegemony to Empire, the Development of the Roman Imperium in the East from 148 to 62 B.C.* (California, 1995)

J.F. Lazenby, *Hannibal's War* (Warminster, 1978)

K. Lehmann-Hartleben, *Die antiken Hafenanlagen des Mittelmeeres* (Leipzig, 1923)

K. Lomas, *Rome and the Western Greeks, Conquest and Acculturation in Southern Italy* (London, 1993)

A.H. McDonald and F.W. Walbank, 'The Treaty of Apamea (188 B.C.): the Naval Clauses', *JRS* 59 (1969), 30–39

E. Maroti, 'Diodotos Tryphon et la Piraterie', *Acta Antiqua* 10 (1962), 187–194

R. Meiggs, *Roman Ostia* (Oxford, 1960)

R. Meiggs, *Trees and Timber in the Ancient World* (Oxford, 1983)

F. Meijer, *A History of Seafaring in the Classical World* (London, 1986)

J.S. Morrison, 'Hyperesia in Naval Contexts', *JHS* 104 (1984), 48–59

J.S. Morrison, 'Athenian Sea Power in 323/2 B.C.: Dream and Reality', *JHS* 107 (1987), 88–97

J.S. Morrison and R.T. Williams, *Greek Oared Ships, 900–322 B.C.* (Cambridge, 1968)

J.S. Morrison and J.F. Coates, *The Athenian Trireme* (Oxford, 1986)

W.M. Murray, 'A trireme named *Isis*: the *graffito* from Nymphaion', *INJA* 30 (2001), 250–256

H.A. Ormerod, *Piracy in the Ancient World* (Liverpool, 1979)

G.C. and C. Picard, *Carthage* (London, 1968)

F. Piejko, 'Episodes from the Third Syrian War in a Gurob Papyrus, 246 B.C.', *Archiv für Papyrusforschung* 36 (1990), 13–27

A. Poidebard and J. Lauffray, *Sidon, Amenagements antiques du port de Saida* (Beirut, 1951)

J.H. Pryor, *Geography, Technology and War, Studies in the Maritime History of the Mediterranean, 649–1571* (Cambridge, 1988)

N.K. Rauh, *Merchants, Sailors and Pirates in the Roman World* (Stroud, 2003)

H.R. Rawlings, 'Antiochus the Great and Rhodes, 197–191 B.C.', *AJAH* 1 (1976), 2–28

G. Reger, 'The Political History of the Kyklades, 260–200 B.C., *Historia* 43 (1994), 32–69

G. Reger, 'The Date of the Battle of Kos', *AJAH* 10 (1985), 155–177

W. Rodgers, *Greek and Roman Naval Warfare* (Annapolis, 1937)

H.H. Scullard, *The Elephant in the Greek and Roman World* (London, 1974)

R. Seager, *Pompey, a Political Biography* (Oxford, 1979)

J. Seibert, 'Philokles, Sohn des Apollodoros, Konig der Sidonier', *Historia* 19 (1970), 337–351

N.V. Sekunda, 'Athenian Demography and Military Strength, 338–322 B.C.', *ABSA* 61 (1997), 340–355

R.K. Sherk, *Rome and the Greek East to the Death of Augustus* (Cambridge, 1984)

A.N. Sherwin-White, *Roman Foreign Policy in the East, 168 B.C. to A.D. 1* (Oklahoma, 1983)

A.W. Sleeswyk and F. Meijer, 'Launching Philopator's 'forty'', *IJNA* 23 (1994), 115–118

P. de Souza, 'Romans and Pirates in a Late Hellenistic Oracle from Pamphylia', *Classical Quarterly* 47 (1997), 477–481

P. de Souza, *Piracy in the Greco-Roman World* (Cambridge, 1999)

S. Spyridakis, *Ptolemaic Itanos and Hellenistic Crete* (California, 1970)

C.G. Starr, *The Roman Imperial Navy 31 B.C.–A.D. 324* (Cambridge, 1960)

C.G. Starr, *The Influence of Sea Power on Ancient History* (Oxford, 1989)

R.H.A. Talbert, *Timoleon and the revival of Greek Sicily, 344–317 B.C.* (Cambridge, 1974)

W.W. Tarn, 'The Dedicated Ship of Antigonos Gonatas', *JHS* 30 (1910), 209–222

W.W. Tarn, 'The Fleets of the First Punic War', *JHS* 27 (1907), 48–60

W.W. Tarn, *Antigonos Gonatas* (Oxford, 1913)

W.W. Tarn, *Hellenistic Military and Naval Developments* (Cambridge, 1930)

H.J.W. Tillyard, *Agathocles* (Cambridge, 1908)

G.K. Tipps, 'The Battle of Ecnomus', *Historia* 34 (1985), 432–465

E. Van't Dack and H. Hauben, 'L'Apport Egyptien a l'Armee navale Lagide', in H. Maehler and V.M. Strocka, *Das Ptolemaische Agyptien* (Mainz, 1978), 59–94

F.W. Walbank, *Philip V of Macedon* (Cambridge, 1940)

F.W. Walbank, 'Sea Power and the Antigonids', in W.L. Adams and E.N. Borza, *Philip II, Alexander the Great and the Macedonian Heritage* (Washington DC, 1982), 23–236

T. Walek, 'Les Operations Navales pendant la Guerre Lamiaque', *Revue de Philologie* 48 (1924), 23–30

H. Wallinga, *The Boarding Bridge of the Romans* (Groningen, 1956)

B.H. Warmington, *Carthage* (London, 1960)

Index